Diffusion Source Localization in Large Networks

Synthesis Lectures on Communication Networks

Editor
R. Srikant, *University of Illinois at Urbana-Champaign*

Founding Editor Emeritus
Jean Walrand, *University of California, Berkeley*

Synthesis Lectures on Communication Networks is an ongoing series of 75- to 150-page publications on topics on the design, implementation, and management of communication networks. Each lecture is a self-contained presentation of one topic by a leading expert. The topics range from algorithms to hardware implementations and cover a broad spectrum of issues from security to multiple-access protocols. The series addresses technologies from sensor networks to reconfigurable optical networks.
The series is designed to:

- Provide the best available presentations of important aspects of communication networks.

- Help engineers and advanced students keep up with recent developments in a rapidly evolving technology.

- Facilitate the development of courses in this field

Diffusion Source Localization in Large Networks
Lei Ying and Kai Zhu
2018

Communications Networks: A Concise Introduction, Second Edition
Jean Walrand and Shyam Parekh
2017

BATS Codes: Theory and Practice
Shenghao Yang and Raymond W. Yeung
2017

Analytical Methods for Network Congestion Control
Steven H. Low
2017

Diffusion Source Localization in Large Networks

Lei Ying and Kai Zhu

ISBN: 978-3-031-79284-7   paperback
ISBN: 978-3-031-79285-4   ebook
ISBN: 978-3-031-79286-1   hardcover

DOI 10.1007/978-3-031-79285-4

A Publication in the Springer series
SYNTHESIS LECTURES ON COMMUNICATION NETWORKS

Lecture #21
Series Editor: R. Srikant, *University of Illinois at Urbana-Champaign*
Founding Editor Emeritus: Jean Walrand, *University of California, Berkeley*
Series ISSN
Print 1935-4185   Electronic 1935-4193

Diffusion Source Localization in Large Networks

Lei Ying and Kai Zhu
Arizona State University

SYNTHESIS LECTURES ON COMMUNICATION NETWORKS #21

ABSTRACT

Diffusion processes in large networks have been used to model many real-world phenomena, including how rumors spread on the Internet, epidemics among human beings, emotional contagion through social networks, and even gene regulatory processes. Fundamental estimation principles and efficient algorithms for locating diffusion sources can answer a wide range of important questions, such as identifying the source of a widely spread rumor on online social networks. This book provides an overview of recent progress on source localization in large networks, focusing on theoretical principles and fundamental limits. The book covers both discrete-time diffusion models and continuous-time diffusion models. For discrete-time diffusion models, the book focuses on the Jordan infection center; for continuous-time diffusion models, it focuses on the rumor center. Most theoretical results on source localization are based on these two types of estimators or their variants. This book also includes algorithms that leverage partial-time information for source localization and a brief discussion of interesting unresolved problems in this area.

KEYWORDS

source localization, diffusion processes, epidemic processes, Jordan infection center, rumor center, large networks, social networks

To Ethan, Lingfang, and my parents—L.Y.
To Tu, my parents, and grandparents—K.Z.

Contents

Preface

We are living in an ever-connected world because of the proliferation of the Internet, mobile devices, and online social networks. Mathematical modeling and analysis of diffusion processes in large networks can help us understand many of the phenomena and problems in this networked world—such as epidemic threshold (i.e. the condition under which an epidemic will break out) and influence minimization/maximization on social networks—and have therefore become a research topic of great interest in recent years. This book studies retrospective analysis of diffusion in networks, in particular the question of how to find the source or sources of a diffusion process given incomplete information on both the network and diffusion. This book is a result of exciting research by the authors since 2011. It is suited for graduate students and researchers who are interested in this problem of large network diffusion. The goal of the book is to provide an overview of key theoretical discoveries on this topic. We feel it is more important for readers to understand the intuition behind the results than to know the detailed proofs that, in some cases, can be quite tedious. Therefore, throughout this book, we use examples to explain the analysis and present simplified proofs based on additional assumptions.

ORGANIZATION OF THE BOOK

We start this book by discussing the motivation and background of the source localization problem in Chapter 1. Chapter 2 focuses on discrete-time diffusion models and source estimators based on the Jordan infection center. Starting from tree networks and independent cascade models, we introduce the short-fat tree (SFT) algorithm based on the Jordan infection center and show that the estimator identified by the SFT is the maximum likelihood (ML) estimator of the diffusion source. Then, we consider the Erdos-Renyi random graph and present conditions under which the source is the unique Jordan infection center with high probability. We further introduce NETSLEUTH, which is a heuristic source localization algorithm for general networks.

Chapter 3 introduces rumor centrality and rumor center for continuous-time diffusion models. We first prove that the rumor center is the ML estimator of the diffusion source on regular tree networks, and then present source localization algorithms based on rumor centrality for general networks. Fundamental limits on the detection probability of rumor center are also included.

In Chapter 4, we introduce algorithms that use partial timestamps—the times at which nodes were infected—for locating the source. Besides providing a source estimator, these algorithms rank the nodes according to their likelihood of being the source, which can be very useful in practice. Finally, in Chapter 5, we briefly discuss several unresolved problems in the field.

Lei Ying and Kai Zhu
June 2018

Acknowledgments

The authors would like to thank Prof. R. Srikant for inviting us to write this book.

Lei Ying and Kai Zhu
June 2018

CHAPTER 1

Motivation and Background

Diffusion processes have been used to model many real-world phenomena, including rumor spreading on the Internet, epidemics in human beings, emotional contagion through social networks, and even gene regulatory processes. Diffusion source localization is to identify the source(s) of a diffusion process based on observations such as the states of the nodes and a subset of timestamps at which the diffusion process reached the corresponding nodes. For example, Figure 1.1 is a snapshot of an Erdos-Renyi (ER) random graph [Erdos and Renyi, 1959] where the red nodes are infected nodes under an epidemic process following the independent cascade (IC) model [Goldenberg et al., 2001]. The source localization problem in this case is to identify "patient-zero" (the boxed node) according to this snapshot.

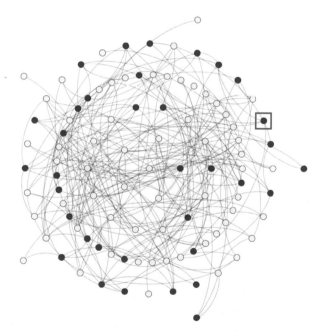

Figure 1.1: A snapshot of an ER random graph under an epidemic process.

The solutions to this problem can answer a wide range of important questions and have significant societal and economic impacts. For example, epidemic diseases are great threats to global health. The 2009 H1N1 virus alone resulted in 151,700 to 575,400 deaths globally [h1n].

Locating an epidemic source can help identify the transmission media of the disease. For a computer virus spreading on the Internet, tracing the source helps locate the virus creator. For news over social media networks, locating the sources helps users verify the credibility of the news. Recently, source localization has also been proposed as a technique for identifying the infusion hubs of human diseases [Feizi et al., 2014] in human gene regulatory networks. If successful, it can lead to new knowledge and new treatments of human diseases.

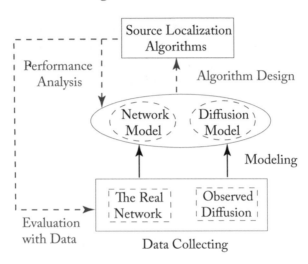

Figure 1.2: The key steps involved in diffusion source localization.

Solving the source localization problem in an analytical manner is a complex process, as shown in Figure 1.2. First, a network model, which is mathematically trackable, needs to be developed to represent the real network. Second, an analytical model is needed to describe the diffusion process. Third, given the network and diffusion models, source localization algorithms need to be developed. Finally, the models and the algorithms need to be refined based on both theoretical performance analysis and evaluation using real data sets. We note that both the network modeling and diffusion modeling have been at the core of network science research, independent from being key steps for identifying diffusion sources. Significant progress has been made in both areas (see [Barabasi and Albert, 1999, Chen et al., 2009, Eguiluz and Klemm, 2002, Erdos and Renyi, 1959, 1960, Ganesh et al., 2005, Gruhl et al., 2004, Kempe et al., 2003, 2005, Kephart and White, 1991, Medina et al., 2000, Moore and Newman, 2000, Newman, 2002, Newman and Watts, 1999, Newman et al., 2002, Nickel, 2006, Pastor-Satorras and Vespignani, 2001, Watts and Strogatz, 1998] and references within). The focus of this book is to introduce recent developments of source localization algorithms under different types of networks, diffusion and information models, involving the three dashed arrow lines in Figure 1.2.

We will start from discrete-time diffusion models and source localization algorithms based on Jordan infection centers in Chapter 2 and discuss the results for continuous-time diffusion models such as rumor centrality in Chapter 3. We will then introduce algorithms that exploit partial timestamps in Chapter 4. Finally, we conclude with open questions in Chapter 5.

CHAPTER 2

Source Localization under Discrete-Time Diffusion Models

This chapter focuses on source localization under discrete-time diffusion models. We will start from simple diffusion and network models and then extend the results to more general models.

2.1 THE INDEPENDENT CASCADE (IC) DIFFUSION MODEL ON TREE NETWORKS

2.1.1 MODELS

- Network model: We consider a tree network, denoted by g, which is an undirected graph and in which any two nodes are connected by a unique path.

- Diffusion model: We consider the IC model [Goldenberg et al., 2001] for information diffusion and assume a time-slotted system. Each node has two possible states: active (or called infected) and inactive (or called susceptible). At time slot 0, all nodes are inactive except the source. At the beginning of each time slot, an active node, which was activated in the previous time slot, attempts to activate its inactive neighbors. If an attempt is successful, the corresponding node becomes active at the next time slot; otherwise, the node remains inactive. The weight of edge (u, v) represents the probability that node u successfully activates node v, called the *infection probability* of the edge. Whether an attempt is successful or not is independent of any other attempts. An active node only attempts to activate each of its inactive neighbors once. Denote by q_{uv} the infection probability of edge (u, v). We assume $q_{uv} = q_{vu}$ throughout the chapter since the graph is undirected.

- Information model: We assume that a complete snapshot $\mathcal{O} = \{\mathcal{I}, \mathcal{H}\}$ of the network at time t (called the *observation time*) is given, where $\mathcal{I}$ is the set of active nodes and $\mathcal{H}$ is the set of inactive nodes. For simplicity, we assume the observation time t follows a geometric distribution with parameter α. In other words, at the end of each time slot, a snapshot is taken with probability α if the snapshot has not been taken so far.

Example

Figure 2.1 shows a simple example of diffusion under the IC model. Nodes in red are active and nodes in white are inactive. At time slot 0, source s is the only active node. At the beginning of time slot 1, s attempts to activate its neighbors (the red arrows represent the attempts). As shown in Figure 2.1, node a and b are successfully activated. At the beginning of time slot 2, a and b attempt to activate their inactive neighbors, and c and d are successfully activated. Based on the information model in this chapter, if the observation time is 2, the snapshot we have includes the set of active nodes $\mathcal{I} = \{s, a, b, c, d\}$ at time slot 2.

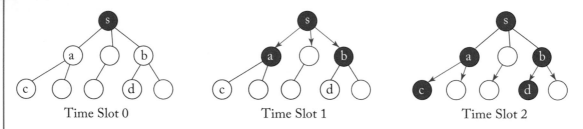

Figure 2.1: An example of the IC diffusion model.

2.1.2 PROBLEM FORMULATION

Given an undirected graph g, denote by $\mathcal{E}(g)$ the set of edges in g and denote by $\mathcal{V}(g)$ the set of nodes in g. Based on $\mathcal{O}$, the goal is to identify the diffusion source. We further assume observation time t is unknown. The problem can be formulated as the maximum likelihood (ML) problem:

$$\arg\max_{v \in \mathcal{V}(g)} \Pr(\mathcal{O}|v), \tag{2.1}$$

where $\Pr(\mathcal{O}|v)$ is the probability to have snapshot $\mathcal{O}$ if node v is the source. The infected nodes form a connected component under the IC model, called the *infection subgraph* and denoted by g_i. Since the source must be an infected node, the ML problem can be simplified to

$$\arg\max_{v \in \mathcal{I}} \Pr(\mathcal{O}|v),$$

and the search of the diffusion source can be restricted to the infection subgraph. We assume the observation time t, which itself is a random variable, is independent of the source node.

2.1.3 THE SHORT-FAT TREE ALGORITHM

In this section, we first present the Short-Fat-Tree (SFT) algorithm developed by Zhu and Ying [2016a], which outputs the ML estimator for tree networks. We first introduce some notation and definitions.

- d_{uv} : the hop distance between two nodes.

- $e(v, \mathcal{I})$: the *infection eccentricity* of infected node v, which is the maximum hop distance from the node to all infected nodes, i.e.,

$$e(v, \mathcal{I}) = \max_{u \in \mathcal{I}} d_{vu}.$$

- The *Jordan infection centers* of a graph are the nodes with the minimum infection eccentricity [Zhu and Ying, 2016a]. Let e^* denote the infection eccentricity of a Jordan infection center.

- Given node v and the set of infected nodes $\mathcal{I}$, the set of *boundary nodes* with respect to node v is defined as

$$\mathcal{B}(v, \mathcal{I}) = \{w \in \mathcal{I} | d_{vw} = e(v, \mathcal{I})\},$$

which is the set of infected nodes that are furthest away from node v.

- The weighted boundary node degree (WBND) with respect to node v and the set of infected nodes $\mathcal{I}$ is defined as

$$\sum_{(u,w) \in \mathcal{E}: u \in \mathcal{B}(v, \mathcal{I}), w \notin \mathcal{I}} |\log(1 - q_{uw})|. \qquad (2.2)$$

Note that edge (u, w) is an edge that connects a boundary node and a healthy node. $|\log(1 - q_{uw})|$ is the weight associated with edge (u, w) in WBND, which is different from q_{uw}.

Example

Figure 2.2 shows a simple example, where nodes in red are infected nodes and nodes in white are healthy nodes. The infection eccentricity of node s is 2 because it can reach all infected nodes with at most two hops. The infection eccentricity of other infected nodes is at least 3. Therefore, s is the unique Jordan infection center in this example. Note that

$$d_{sa} = d_{sb} = d_{sc} = e(s, \mathcal{I}) = 2,$$

so nodes a, b, and c are the boundary nodes with respect to node s, i.e.,

$$\mathcal{B}(s, \mathcal{I}) = \{a, b, c\}.$$

We next introduce the SFT algorithm (Algorithm 2.1) based on the concepts and notation defined above. The algorithm identifies the source by finding the Jordan infection center with the largest WBND. It is called the Short-Fat-Tree algorithm because the infection subtree rooted at the node selected by SFT is the shortest among all infection subtrees (i.e., short), and has the maximum WBND among all shortest infection subtrees (i.e., fat). The pseudo code is presented in Algorithm 2.1. The implementation of SFT first lets infected nodes broadcast their IDs in the network, and the nodes that first receive the IDs of all infected nodes are the Jordan infection centers. Then among all Jordan infection centers, SFT selects the one with the largest WBND.

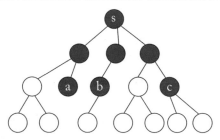

Figure 2.2: An example of Jordan infection center and boundary nodes.

Example

A simple example is presented in Figure 2.3 to illustrate the SFT algorithm. Each node has a unique node ID. The red nodes are infected and the white nodes are healthy. For simplicity, we assume the weights of all edges are equal to $|\log(0.5)|$. The vector next to each infected node records the distance from it to all infected nodes. Initially at Iteration 0, each infected node only knows the distance to itself. For example, $[0***]$ next to node 1 means that the distance from node 1 to itself is 0 and the distance from node 1 to other nodes is unknown. At Iteration 1, each infected node broadcasts its ID to its neighbors. Upon receiving the IDs from node 1 and node 5, node 2 updates its vector to $[1\ 0\ *\ *\ 1]$ and broadcasts node 1 and node 5's IDs to its neighbors. The figure in the middle shows the updated vectors after all ID exchanges are finished at Iteration 1. At Iteration 2, node 2 receives the IDs of nodes 3 and 4 from node 1 and updates its vector to $[1\ 0\ 2\ 2\ 1]$. Note that at Iteration 3, nodes 1 and 2 do not receive new IDs. Therefore, node 1 and node 2 report themselves as the Jordan infection centers which are circled in Figure 2.3. The boundary node with respect to node 1 is 5. The WBND of node 1 is 0 since node 5 does not have a healthy neighbor. Similarly, the boundary nodes with respect to node 2 are nodes 3 and 4 and the WBND is $4|\log(0.5)|$. Therefore, node 2 has larger WBND and is chosen to be the estimator of the diffusion source.

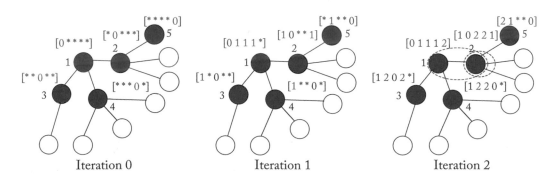

Figure 2.3: An example of the Short-Fat-Tree algorithm.

Algorithm 2.1 The Short-Fat Tree Algorithm

Input : $\mathcal{I}, g$;
Output : $v^\dagger$ (the estimator of diffusion source)

1: Set subgraph g_i to be the subgraph of g induced by node set $\mathcal{I}$.
2: **for** Each $v \in \mathcal{I}$ **do**
3:     Initialize an empty dictionary D_v for node v.
4:     Set $D_v[v] = 0$ and $D_v[u] = *$ for $u \neq v$.
5: **end for**
6: Each node receives its own node ID at time slot 0.
7: Set time slot $t = 1$.
8: **while** No node receives $|\mathcal{I}|$ distinct node IDs **do**
9:     **for** $v \in \mathcal{I}$ **do**
10:         **if** Node v received new node IDs in $t - 1$ time slot, where a "new" ID is an ID that node v had not received before time slot $t - 1$ **then**
11:             Node v broadcasts the new node IDs to its neighbors in g_i.
12:         **end if**
13:     **end for**
14:     **for** $v \in \mathcal{I}$ **do**
15:         **if** Node v receives a new node ID u which is not in D_v. **then**
16:             Set $D_v[u] = t$.
17:         **end if**
18:     **end for**
19:     $t = t + 1$.
20: **end while**
21: Set $\mathcal{S}$ to be the set of nodes that receive $|\mathcal{I}|$ distinct node IDs.
22: **for** $v \in \mathcal{S}$ **do**
23:     Compute WBND of v using Algorithm 2.2.
24: **end for**
25: **return** $v^\dagger \in \mathcal{S}$ with the maximum WBND.

Remark: Note Equation (2.2) requires the infection probabilities of all edges in the network, which could be hard to obtain in practice. When the infection probabilities are not available, we can assume each edge has the same infection probability q and WBND becomes

$$\left(\sum_{u \in \mathcal{B}(v, \mathcal{I})} \deg(u) - |\mathcal{B}(v, \mathcal{I})| \right) |\log(1 - q)|,$$

Algorithm 2.2 The WBND Algorithm

Input : v, D_v (Dictionary of distance from v to other nodes), g, $\mathcal{I}$, t;
Output : WBND(v)

1: Set $\mathcal{B}$ to be empty.
2: **for** u in the keys of D_v **do**
3:    **if** $D_v[u] = t$ **then**
4:       Add u to $\mathcal{B}$.
5:    **end if**
6: **end for**
7: Set $x = 0$;
8: **for** $w \in \mathcal{B}$ **do**
9:    Find the neighbor u of w such that $D_v[u] = t - 1$.
10:    Set $x = x - |\log(1 - q_{wu})| + \sum_{y \in \text{neighbors}(w)} |\log(1 - q_{wy})|$.
11: **end for**
12: **return** x.

---

where $\deg(u)$ is the degree of node u.

We can then define the boundary node degree (BND) of node v to be

$$\sum_{u \in \mathcal{B}(v, \mathcal{I})} \deg(u) - |\mathcal{B}(v, \mathcal{I})|, \tag{2.3}$$

which is only related to the degree of a boundary node and can be used to replace WBND as the tie-breaking among Jordan infection centers in SFT when the infection probabilities are unknown. Our numerical examples will show that the performance of SFT using BND or WBND is similar. To differentiate the two algorithms, we name the algorithm which uses WBND *wSFT* and the one which uses BND *SFT*.

2.1.4   THE ML ESTIMATOR

We now analyze SFT under the IC model using the live-edge model [Kempe et al., 2003]. We first explain the intuition by assuming the infection probability is q for all edges. The result is proved for the general case. Under the live-edge model, each existing edge in the tree network is a live edge with probability q and a dead edge with probability $1 - q$. Figure 2.4 shows a simple example of the live-edge graph model, where the edges marked by "x" are dead edges. After removing the dead edges, we obtain the graph on the right-hand-side.

Note that the IC model and live-edge graph model are closely related. In the IC model, after u becomes infected, it attempts to infect its neighbor w with probability q once. Therefore,

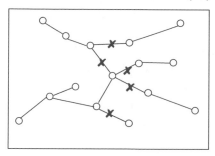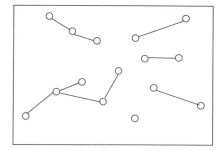

Figure 2.4: An example of the live-edge model.

we can assume that a biased coin with parameter q is flipped for edge (u, w) when u tries to infect w in the IC model. Note that the probability that node w is infected by node u remains the same whether the coin is flipped at the moment when node u attempts to infect w or prior to that but is revealed when the attempt occurs. Therefore, we can assume the coins are flipped before the diffusion process starts. When one node attempts to infect one of its neighbors, we check the stored coin flipping realization to determine whether the infection succeeds. This process then coincides with the live-edge model since we only change the time of the coin flipping. Under the live edge model, the diffusion process includes two steps. First, each edge (u, w) flips a biased coin and becomes a live edge with probability q. Then, the infection spreads over all live edges deterministically, starting from the source. From the discussion above, we can see that a diffusion trace with duration t must be the t-hop neighborhood of the source in at least one live-edge graph. For example, a diffusion trace on the left-hand side of Figure 2.5 coincides with the two-hop neighborhood of the live-edge graph on the right-hand side.

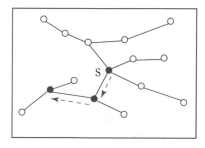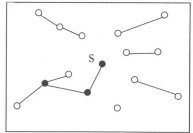

Figure 2.5: Each diffusion trace with duration of t time slots must coincide with the t-hop neighborhood of at least one live-edge graph.

Since a diffusion trace can be equivalently generated (with the same probability) using the live-edge graph model by first generating the corresponding live-edge graph and then finding the t-hop neighborhood of the source, the probability of having observation $\mathcal{O}$ is equal to the total probability of those live-edge graphs on which the trace coincides with the t-hop neighborhood of the source.

Let $\mathcal{K}$ denote the set of all live-edge graphs for graph g and with parameter q. Given observation $\mathcal{O}$ and node v, let $\mathcal{K}_{\mathcal{O},v}$ be the set of live-edge graphs such that for each of them, there exists a positive integer d such that $\mathcal{I}$ coincides with the d-hop neighborhood of v on the live-edge graph. For example, given the observation as shown in Figure 2.6, live-edge graph 1 is in set $\mathcal{K}_{\mathcal{O},v}$ while live-edge graph 2 is not because the 2-hop neighborhood of v in live-edge graph 2 includes two healthy nodes.

	Live-Edge Graph 1	Live-Edge Graph 2
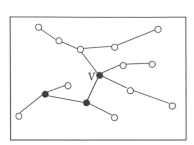	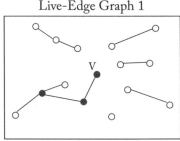	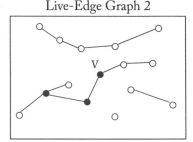

Figure 2.6: Live-edge graph 1 is in $\mathcal{K}_{\mathcal{O},v}$ but live-edge graph 2 is not.

Given a live-edge graph $k \in \mathcal{K}$ and letting $\mathcal{E}_k$ denote the set of live edges in k, then the probability of having k is

$$p_k = q^{|\mathcal{E}_k|}(1-q)^{|\mathcal{E}\setminus\mathcal{E}_k|}.$$

Let $\mathcal{N}(k,v,t)$ denote the t-hop neighborhood of node v in the live-edge graph k, including node v itself. The probability of having observation $\mathcal{O}$, conditioned on v as the source is the total probability of all live-edge graphs in set $\mathcal{K}_{\mathcal{O},v}$, which is

$$\Pr(\mathcal{O}|v) = \sum_{k\in\mathcal{K}_{\mathcal{O},v}} p_k \left(\sum_{t=1}^{\infty}(1-\alpha)^{t-1}\alpha 1_{\mathcal{N}(k,t,v)=\mathcal{O}}\right). \tag{2.4}$$

Equation (2.4) establishes a fundamental connection between the conditional probability $\Pr(\mathcal{O}|v)$ and the number of live-edge graphs in $\mathcal{K}(\mathcal{O}, v)$. However, we have $2^{|\mathcal{E}|}$ live-edge graphs in $\mathcal{K}$, so it is computationally impossible to enumerate all live-edge graphs in $\mathcal{K}$ to calculate the total probability. Also for two non-neighboring nodes v and w, it is difficult even to compare $\mathcal{K}_{\mathcal{O},v}$ and $\mathcal{K}_{\mathcal{O},w}$ because the live-edge graphs in the two sets are quite different. However, if v and w are neighbors on g, then we have the following lemma, which we call neighboring-nodes lemma. Note that all the following results are proved with heterogeneous infection probabilities.

Lemma 2.1   Neighboring-nodes lemma *Consider two neighboring nodes u and v. If $e(v, \mathcal{I}) > e(u, \mathcal{I})$, then*

$$\mathcal{K}_{\mathcal{O},v} \subset \mathcal{K}_{\mathcal{O},u}.$$

In other words, if u has a smaller infection eccentricity, then the set of live-edge graphs that can lead to observation $\mathcal{O}$ with node v as the source is a (strict) subset of the set of live-edge graphs that can lead to observation $\mathcal{O}$ with node u as the source.

Proof. We will prove that any live-edge graph that belongs to $\mathcal{K}_{\mathcal{O},v}$ also belongs to $\mathcal{K}_{\mathcal{O},u}$. If a live-edge graph k belongs to $\mathcal{K}_{\mathcal{O},v}$ for some v, then

$$\mathcal{I} = \mathcal{N}(k, v, e(v, \mathcal{I})).$$

This is because node v can reach all infected nodes within $e(v, \mathcal{I})$ hops according to the definition of infection eccentricity and node v should not be able to reach any healthy node within $e(v, \mathcal{I})$ hops when $k \in \mathcal{K}_{\mathcal{O},v}$.

Since nodes u and v are neighbors and $e(v, \mathcal{I}) > e(u, \mathcal{I})$,

$$\mathcal{N}(k, u, e(u, \mathcal{I})) \subseteq \mathcal{N}(k, v, e(v, \mathcal{I})) = \mathcal{I}.$$

On the other hand,

$$\mathcal{I} \subseteq \mathcal{N}(k, u, e(u, \mathcal{I}))$$

because node u can reach all infected nodes within $e(u, \mathcal{I})$ hops according to the definition of eccentricity. Therefore, we conclude that

$$\mathcal{N}(k, u, e(u, \mathcal{I})) = \mathcal{I}.$$

and $k \in \mathcal{K}_{\mathcal{O},u}$. $\qquad\square$

Consider the example in Figure 2.6, where the infection eccentricity of node v is 2 and the infection eccentricity of node u is 1. In live-edge graph 1, the set of infected nodes $\mathcal{I}$ coincides with the two-hop neighborhood of v and the one-hop neighborhood of u. Therefore, live-edge graph 1 belongs to both $\mathcal{K}_{\mathcal{O},v}$ and $\mathcal{K}_{\mathcal{O},u}$. On the other hand, for live-edge graph 2, the set of infected nodes coincides with the one-hop neighborhood of u but is a subset of two-hop neighborhood of v, so live-edge graph 2 belongs to $\mathcal{K}_{\mathcal{O},u}$ but does not belong to $\mathcal{K}_{\mathcal{O},v}$.

The lemma above states that for two neighboring nodes u and v such that u has a smaller infection eccentricity, the set of live-edge graphs that are feasible for u includes all live-edge graphs that are feasible for v. This lemma immediately leads to the conclusion that the probability to have the observed snapshot is higher when u is the source than that when v is the source, i.e.,

$$\Pr(\mathcal{O}|u) > \Pr(\mathcal{O}|v).$$

Lemma 2.2   *There exist at most two Jordan infection centers in a tree network. If the tree network has two Jordan infection centers, then the two must be neighbors.* $\qquad\square$

Intuition: If two Jordan infection centers are not adjacent, we pick a node on the path between two Jordan centers and adjacent to one of the Jordan infection centers. It can be shown that the selected node has a smaller infection eccentricity which is a contradiction. If any two Jordan infection centers are adjacent, more than two Jordan infection centers would form a clique which contradicts with the fact that the network is a tree. Therefore, we have at most two Jordan infection centers in a tree network and they are adjacent.

Proof. We first prove that if there are more than one Jordan infection centers, they must be neighbors. Suppose nodes u and v are two Jordan infection centers and $e(v, \mathcal{I}) = e(u, \mathcal{I}) = e^*$. Suppose v and u are not adjacent nodes, i.e., $d(v, u) > 1$. Then, we select a node, say node w, on the path between u and v and is a neighbor of node u as shown in Figure 2.7.

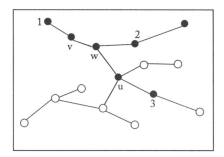

Figure 2.7: An example of Jordan infection centers.

We now study the distance between node w and an infected node a by considering the following two cases. First, suppose the path between node a and node u goes through node w, such as node 1 and node 2 in Figure 2.7. In this case

$$d(w, a) = d(u, a) - 1 = e^* - 1.$$

Now suppose that the path between node a and node u does not pass through node w, such as node 3 in Figure 2.7. Then the path from node a to node v has to pass through node w because node w is on the unique path between u and v. Therefore,

$$d(v, w) + d(w, a) = d(v, a),$$

i.e.,

$$d(w, a) = d(v, a) - d(v, w) \leq e^* - 1.$$

Hence, we conclude that $d(w, \mathcal{I}) = e^* - 1$, which contradicts the fact that the eccentricity of Jordan infection center is e^*.

Therefore all Jordan infection centers must be adjacent to each other. Now suppose there are n Jordan infection centers and $n > 2$. These Jordan infection centers are neighbors to each other so they would form a clique, which contradicts the fact that the graph is a tree.

In summary, there exist at most two Jordan infection centers in tree networks, and if more than one Jordan infection center exists, they must be neighbors.   □

The next lemma shows that the infection eccentricity strictly decreases along the path from a node to the closer Jordan infection center.

Lemma 2.3   *Let c_a denote the Jordan infection center closer to node a and $\mathcal{P}_{a,c_a}$ denote the path from node a to c_a. For any node on path $\mathcal{P}_{a,c_a}$, say node b, we have*

$$e(a,\mathcal{I}) = e(c_a,\mathcal{I}) + d(a,c_a).$$

Proof. Consider the path from node a to the closer Jordan infection center as shown in Figure 2.8, and node b on the path that is the neighbor of c_a. Since node b is not a Jordan infection center, as shown in Figure 2.8, $e(b,\mathcal{I}) = e^* + 1$ and there exists node b' $d(b,b') = e^* + 1$, and b can only reach b' via c_a. As we can see from the figure, the path from node a to node b' has to go through node c_a as well. Therefore,

$$e(a,\mathcal{I}) \geq d(a,b') = d(a,b) + d(b,b') = d(a,c_a) + e^*.$$

On the other hand,

$$e(a,\mathcal{I}) \leq d(a,c_a) + e(c_a,\mathcal{I}) = d(a,c_a) + e^*$$

because node a can reach all infected nodes by first going to node c_a with $d(a,c_a)$ hops and then all infected nodes with at most e^* hops.

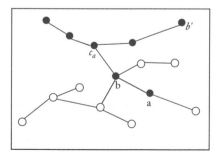

Figure 2.8: An example for illustrating the proof of Lemma 2.3.

□

From the lemmas above, we can conclude the following theorem.

Theorem 2.4   The Jordan center with larger WBND is a solution to (2.1), i.e., an ML estimator of the source.

Proof. Recall $\mathcal{J}$ is the set of Jordan infection centers which includes at most two nodes. We will show that the one with larger WBND is the ML estimator. Define an edge set

$$\mathcal{F} = \{(v, w)|(v, w) \in \mathcal{E}(g), v \in \mathcal{I}, w \in \mathcal{H}\}.$$

We call an edge in $\mathcal{F}$ a frontier edge because it connects an infected node and a healthy node. Define another edge set

$$\mathcal{L} = \{(v, w)|(v, w) \in \mathcal{E}(g), v, w \in \mathcal{I}\}.$$

The edges in $\mathcal{L}$ are the edges between infected nodes. So edges in $\mathcal{L}$ have to be live edges in any live-edge graph that can lead to observation $\mathcal{O}$. In addition, for any Jordan infection center $c \in \mathcal{J}$, we define

$$\mathcal{D}_c(t_c) = \{(v, w)|(v, w) \in \mathcal{E}(g), v \in \mathcal{I}, w \in \mathcal{H}, d(c, w) \leq t_c\},$$

which is set of edges that must be dead edges in a live-edge graph that can lead to observation $\mathcal{O}$ with node c as the source and t_c as the observation time.

Recall that e^* is the infection eccentricity of Jordan infection centers. If $t_c > e^*$, none of the frontier edges should be a live edge in a feasible live-edge graph to make sure healthy nodes are not infected. Therefore,

$$\mathcal{D}_c(t_c) = \mathcal{F} \quad \forall c \in \mathcal{J}, t_c > e^*,$$

and

$$\Pr(\mathcal{O}|c, t_c) = \prod_{(u,v)\in\mathcal{L}} q_{uv} \prod_{(u,v)\in\mathcal{D}_c(t_c)} (1 - q_{uv}) = \prod_{(u,v)\in\mathcal{L}} q_{uv} \prod_{(u,v)\in\mathcal{F}} (1 - q_{uv}),$$

remains a constant for any $t_c > e^*$ and any $c \in \mathcal{J}$.

Now for $t_c = e^*$ for $c \in \mathcal{J}$, we have

$$\Pr(\mathcal{O}|c, e^*) = \prod_{(u,v)\in\mathcal{L}} q_{uv} \prod_{(u,v)\in\mathcal{D}_c(e^*)} (1 - q_{uv}) = \frac{\prod_{(u,v)\in\mathcal{L}} q_{uv} \prod_{(u,v)\in\mathcal{F}}(1 - q_{uv})}{\prod_{(u,v)\in\mathcal{F}\backslash\mathcal{D}_c(e^*)}(1 - q_{uv})},$$

where the numerator is a constant independent of Jordan infection center c. For the denominator, note that $\mathcal{F}\backslash\mathcal{D}_c(e^*) = \{(u, v) \in \mathcal{E} : u \in \mathcal{B}(v, \mathcal{I}), w \notin \mathcal{I}\}$, so

$$\prod_{(u,v)\in\mathcal{F}\backslash\mathcal{D}_c(e^*)} (1 - q_{uv}) = e^{-\text{WBND}}$$

according to the definition of WBND. Therefore, the Jordan infection center with larger WBND is more likely to result in observation $\mathcal{O}$ when being chosen as the source, which concludes the theorem.   □

2.2   SINGLE-SOURCE LOCALIZATION ON THE ER RANDOM GRAPH

In this section, we consider the source localization problem in a non-tree network, the ER random graph [Erdos and Renyi, 1959]. We will show that the SFT algorithm provides strong theoretical guarantees for the ER random graph as well.

The Erdos-Renyi (ER) Random Graph [Erdos and Renyi, 1959]: An ER random graph with n nodes and wiring probability p is a random graph in which any two nodes are connected by an edge with probability p, independent of other edges.   □

For simplicity, we again assume the IC model for diffusion. We further assume homogeneous infection probability q. We will focus on the asymptotic performance of SFT and use the following notation: Given non-negative functions $f(n)$ and $g(n)$:

- $f(n) = O(g(n))$ means there exist positive constants c and m such that $f(n) \leq cg(n)$ for all $n \geq m$;

- $f(n) = \Omega(g(n))$ means there exist positive constants c and m such that $f(n) \geq cg(n)$ for all $n \geq m$;

- $f(n) = \Theta(g(n))$ means that both $f(n) = \Omega(g(n))$ and $f(n) = O(g(n))$ hold;

- $f(n) = o(g(n))$ means that $\lim_{n \to \infty} f(n)/g(n) = 0$; and

- $f(n) = \omega(g(n))$ means that $\lim_{n \to \infty} g(n)/f(n) = 0$.

In this section, we will establish the following fundamental results of SFT on the ER random graph.

(1) When the infection duration is $\frac{2}{3} \frac{\log n}{\log np} - \omega(1)$, SFT identifies the source with probability one (w.p.1) asymptotically (as network size increases).

(2) When the infection duration is $\frac{\log n}{\log np} + \omega(1)$, the probability of identifying the source approaches zero asymptotically under *any* source localization algorithm, i.e., it is *impossible* to identify the source with a non-zero probability.

(3) When the infection duration is $\frac{\log n}{\log np} - \omega(1)$ and $p \geq \frac{9}{\delta^2} \frac{\log n}{n}$, asymptotically, at least $1 - \delta$ fraction of the nodes on the BFS tree starting from the source are leaf nodes. Note that this result does not provide a guarantee on the probability of correctly localizing the source but states that the BFS tree starting from the true source is a "fat" tree, which justifies the SFT algorithm, and is confirmed by the simulation results.

The results are summarized in Figure 2.9. For simplicity, we assume t_u *is an integer*. The proofs of these results are very involved and can be found in Zhu and Ying [2015]. In this book, we will only present the key ideas behind the proofs.

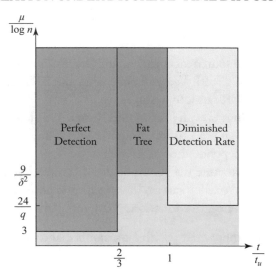

Figure 2.9: Summary of the main results for the ER random graph under the IC model. This figure summarizes the key results in terms of t, the observation time, and $\mu = np$, the average node degree. In the figure, $t_u = \frac{\log n}{\log np}$.

We first discuss the result on probability one detection (result (1)). A well-known result of the ER random graph is that the $\frac{1}{2}\frac{\log n}{\log np}$-hop neighborhood of a node in the ER random graph is a tree with a high probability. Therefore, it is straightforward to show that when the observation time is no longer than $\frac{t_u}{2}$, the actual source is the unique Jordan infection center. The interesting case is when the observation time is larger than $\frac{t_u}{2}$. In such a case, the infection subgraph is not a tree anymore. The following theorem shows that the diffusion source remains the unique Jordan infection center as long as the infection duration is smaller than $\frac{2}{3}t_u$.

Theorem 2.5   Given $p = \Theta\left(\frac{\log n}{n}\right) > 3\frac{\log n}{n}$, and $t = \frac{2}{3}\frac{\log n}{\log(np)} - \omega(1)$, the probability that the source is the only Jordan infection center on the infection subgraph approaches one as $n \to \infty$.
□

The proof of this theorem is very involved. We next present the key ideas. The complete proof can be found in Zhu and Ying [2015].

Back-of-Envelope Proof of Theorem 2.5.   Suppose node s is the diffusion source. We consider the breadth-first-search (BFS) tree in the infection subgraph starting from node s. The BFS tree is constructed by adding nodes to the tree hop-by-hop such that the source is the root node and the k-hop neighbors of node s are at level k of the BFS tree. Note that BFS tree may not be unique because a k-hop neighbor of node s may connect to more than one $(k-1)$-hop neighbors of node s. Our analysis does not depend on which BFS tree is selected if the BFS tree is not unique. Figure 2.10 presents an example of the BSF tree from node s. In the BFS tree, we also added three blue-dotted edges which are on the original graph but not part of the BFS tree. These three edges are named "*Multi–Path (MP) edges*" because the existence of MP edges "destroys" the tree structure and creates multiple paths between a pair of nodes in the ER random graph, which further creates multiple Jordan infection centers. In the example in Figure 2.10, without the MP edges, s is the only Jordan infection center with infection eccentricity of 2. Any other node has infection eccentricity larger than 2. For example, the path for node 6 to reach node 5 without involving the MP edges is $6 - 3 - s - 4 - 5$, which has 4 hops. However, with the MP edges, 5 is a neighbor of 6. In fact, with the MP edges, nodes 1, 3, 4, and 6 all become Jordan infection centers, with infection eccentricity equal to 2.

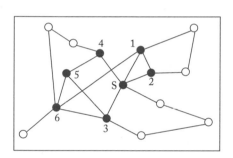
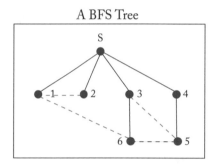

A BFS Tree

Figure 2.10: An example of BFS tree.

We now consider a slightly different graph as shown in Figure 2.11. The infection subgraph still has MP edges. However, we notice that there are no MP edges connecting the blue area and the red area. Because of that, the path between any two nodes that are not in the same area has to go through the root node (source s), and node s remains the only Jordan infection center.

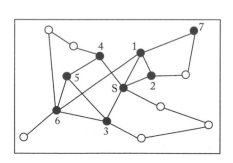
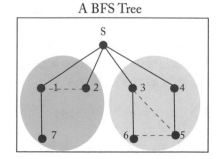

A BFS Tree

Figure 2.11: An example of BFS tree.

An MP edge can bring the leaf nodes of the BFS tree closer to the root of the BFS tree. The key idea of the proof is to show that for any given node, the number of leaf nodes it can reach via MP edges is smaller than the total number of leaf nodes. Therefore, while the MP edges reduce the average distance from a node to the infected nodes, with high probability, the maximum distance, i.e., the infection eccentricity of the node, would remain the same as that in the BFS tree. Therefore, with high probability, the infection eccentricity of source s is t, the observation time, and the infection eccentricity of any other node is larger than t. In this case, source s is the unique Jordan infection center.

We next present a simplified version of the proof using *mean approximation*, where we replace the values of random variables (e.g., node degree) with the means of these random variables. The argument is not rigorous, but nevertheless it includes the key intuition and ideas of the formal proof which repeatedly uses the Chernoff bound to prove that the values of the random variables are close to their means. The complete proof can be found in Zhu and Ying [2015] for those who are interested.

Lemma 2.6   *If there is an MP edge between node a and node b, then the heights of nodes a and b with respect to the BFS tree are either the same or differ by one.*

Proof. Let $h(a)$ denote the height of node a on the BFS tree. According to the way the BFS tree is constructed, $h(a) = d(s, a)$. Without loss of generality, assume $h(b) > h(a)$, i.e., the height of node b is larger than that of node a. Now if there is an edge between node a and b, then

$$h(b) = d(s, b) \leq d(s, a) + d(a, b) = h(a) + 1.$$

So the lemma holds.   □

We now make the following assumption based on the mean approximation.

Regular-BFS-Tree Assumption: We assume the BFS tree from node s is a tree such that each node has $\mu - 1$ offsprings except the root which has μ offsprings, where $\mu = np$.   □

Now under the regular-BFS-tree assumption and mean approximation, we first analyze the structure of the BFS tree, where we further assume that $\frac{\log n}{\log(\mu-1)}$ is an integer. A node is called an MP node if it is an end node of some MP edge. We next calculate the number of MP nodes at each level of the BFS tree.

- Level 0: There is only one node at level 0, which is source s.

- Level 1: There are μ nodes at level 1, each node with $\mu - 1$ offsprings. Consider node a at level 1. Note that node a can have an MP edge with another node at level 1 (there are $\mu - 1$ such nodes), or a node at level 2 which is not an offspring of node a (there are

$(\mu - 1)^2$ such nodes). For the ER random graph, two nodes are connected by an edge with probability μ/n. Therefore, the probability node a is an MP node is

$$1 - \left(1 - \frac{\mu}{n}\right)^{\mu - 1 + (\mu - 1)^2} \approx \frac{\mu^3}{n}. \tag{2.5}$$

Using the mean approximation, we have

$$\mu \left(1 - \left(1 - \frac{\mu}{n}\right)^{\mu - 1 + (\mu - 1)^2}\right) \approx \frac{\mu^4}{n} = n^3 p^4 \approx 0, \tag{2.6}$$

i.e., no MP node at level 1.

- Level 2: Level 2 have $\mu(\mu - 1)$ nodes. Consider node a at level 2. Node a at level 2 can have an MP edge with a level 1 node that is not a's parent, or any other node at level 2, or any node at level 3 that is not a's offspring. Therefore, the probability node a is an MP node is

$$1 - \left(1 - \frac{\mu}{n}\right)^{\mu - 1 + \mu(\mu - 1) - 1 + (\mu - 1)(\mu(\mu - 1) - 1)} \approx \frac{\mu^4}{n}. \tag{2.7}$$

Using the mean approximation, we have

$$\mu(\mu - 1) \left(1 - \left(1 - \frac{\mu}{n}\right)^{\mu - 1 + (\mu - 1)^2}\right) \approx \frac{\mu^6}{n} \approx 0, \tag{2.8}$$

i.e., no MP node at level 2.

- Level $j < \frac{1}{2}\frac{\log n}{\log \mu} - 1$: There are $\mu(\mu - 1)^{j-1}$ nodes at level j. Node a, at level j, is an MP node with probability

$$1 - \left(1 - \frac{\mu}{n}\right)^{\mu(\mu - 1)^{j-2} - 1 + \mu(\mu - 1)^{j-1} - 1 + (\mu - 1)\mu(\mu - 1)^{j-1} - 1} \approx \frac{\mu^{j+2}}{n}, \tag{2.9}$$

where $\mu(\mu - 1)^{j-2} - 1$ is the number of nodes at level $j - 1$ that are not the parent of node a, $\mu(\mu - 1)^{j-1} - 1$ is the number of other nodes at level j, and $(\mu - 1)\mu(\mu - 1)^{j-1} - 1$ is the number of nodes at level $j + 1$ that are not offsprings of node a. Therefore, the expected number of MP nodes at level j is approximately

$$\mu(\mu - 1)^{j-1} \times \frac{\mu^{j+1}}{n} \approx \frac{\mu^{2j+2}}{n} \approx 0. \tag{2.10}$$

- Level $j = \frac{1}{2}\frac{\log n}{\log \mu} - 1$: Use the same analysis above, we have the expected number of MP nodes at level j is approximately

$$\frac{\mu^{2j+2}}{n} \approx 1. \tag{2.11}$$

- Level $j > \frac{1}{2} \frac{\log n}{\log \mu} - 1$: Similarly argument yields that level j has

$$\frac{\mu^{2j+2}}{n} \qquad (2.12)$$

MP nodes.

Let m_j denote the number of MP nodes on level j. From the discussion above, we have

$$m_j = \begin{cases} 0, & j \le \frac{1}{2} \frac{\log n}{\log \mu} - 1 \\ \frac{\mu^{2j+2}}{n}, & \text{otherwise.} \end{cases} \qquad (2.13)$$

Recall that t is the observation time, which is also the depth of the BFS tree. Now we consider node a $(a \ne s)$ on the BFS tree. Suppose that the infection eccentricity of node a is no larger than t, i.e., $e(a, \mathcal{I}) \le t$. Let $\mathcal{T}_a$ denote the subtree of the BFS tree rooted at node a. Node a can reach the leaf nodes on $\mathcal{T}_a$ with no more than t hops. There are at most $(\mu - 1)^{t-1}$ such leaf nodes on the subtree $\mathcal{T}_a$. For a leaf node not on $\mathcal{T}_a$, say node b, node a can reach node b with no more than t hops *only if* the shortest path between a and b includes at least one MP edge, i.e., one or more MP nodes. Let c be the MP node closest to node b on the shortest path between nodes a and b. Then b must be a leaf node on some subtree $\mathcal{T}_d$ which satisfies the following two conditions:

C1 node d is an ancestor of node c, and

C2 $(l_c - l_d) + (t - l_d) \le t$, where l_c and l_d are the levels of node c and node d, respectively.

Note that the second inequality guarantees that the distance between node c and b is no more than t on the BFS tree. Since node c is the last MP node on the shortest path from node a to node b, on the shortest path, the edges connecting node c and node b are the edges on the BFS tree.

Example

Figure 2.12 illustrates the discussion above with a simple example. Because of the two MP edges (blue dashed edges), node a also becomes a Jordan infection center, besides node s. For node a to reach node b, it has to go through the MP edge (a, c) and the MP node c. Node b is not a leaf node of subtree $\mathcal{T}_c$ but is a leaf node of subtree $\mathcal{T}_d$. Node d is an ancestor of node c. Furthermore, since $t = 4$, $l_c = 3$ and $l_d = 2$. Therefore, we have

$$(l_c - l_d) + (t - l_d) = 3 - 2 + 4 - 2 = 3 < 4.$$

Therefore, node d satisfies both conditions (C1) and (C2).   □

Now if both d and c satisfy the two conditions above and $l_d < l_c$, then $\mathcal{T}_c$ is a subtree of the subtree $\mathcal{T}_d$. See the example in Figure 2.12. In other words, the leaf nodes of $\mathcal{T}_c$ are also leaf

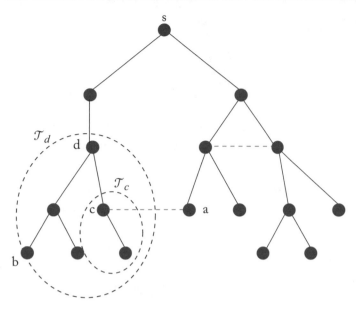

Figure 2.12: An example of the impact of MP nodes.

nodes of $\mathcal{T}_d$. Therefore, to bound the number of leaf nodes that can be reached via node c as the *last* MP node, we just need to find its oldest ancestor that satisfies the two conditions (C1) and (C2), which can be easily verified is at level

$$\left\lceil \frac{l_c}{2} \right\rceil$$

by finding the smallest value that satisfies condition (C2).

Let $\tilde{\mathcal{T}}_c$ denote the subtree rooted at the ancestor of node c at level $\left\lceil \frac{l_c}{2} \right\rceil$. Since the BFS tree is assumed to be a regular tree, $\tilde{\mathcal{T}}_c$, it includes $(\mu - 1)^{t - \left\lceil \frac{l_c}{2} \right\rceil}$ leaf nodes. Let $\mathcal{L}_c$ be the set of leaf nodes of subtree $\tilde{\mathcal{T}}_c$ and $\mathcal{C}$ be the set of MP nodes. Based on the mean approximation of the number of MP nodes on each level, we obtain

$$\left| \bigcup_{c \in \mathcal{C}} \mathcal{L}_c \right| \leq \sum_{j=1}^{t} m_j (\mu - 1)^{t - \left\lceil \frac{j}{2} \right\rceil}$$

$$\approx \sum_{j = \frac{1}{2} \frac{\log n}{\log \mu}}^{t} \frac{\mu^{t + \frac{3}{2} j}}{n}$$

$$\approx \Theta\left(\frac{\mu^{2.5t}}{n}\right).$$

Therefore, when $t = \frac{2}{3}\frac{\log n}{\log(np)} - \omega(1)$, we have

$$\left| \bigcup_{c \in C} \mathcal{L}_c \right| = o(\mu^t). \tag{2.14}$$

Note that the BFS tree has at least $(\mu - 1)^t$ leaf nodes that are not on $\mathcal{T}_a$. Therefore, we can conclude that node a cannot reach all leaf nodes via MP edges, and the distance between node a and some leaf node must be at least $t + 1$. Hence, node a cannot be a Jordan infection center.

$\square$

We next present the threshold on t after which it is impossible for any algorithm to find the actual source with a non-zero probability asymptotically. Recall the ER graph where an edge exists between two nodes with probability p.

Theorem 2.7   If $24 \log n < q\mu << \sqrt{n}$ and $q > 0$ is a constant, then

$$\lim_{n \to \infty} \Pr(\mathcal{I} = \mathcal{V}(g)) = 1$$

when the observation time

$$t = \frac{\log n}{\log(npq)} + \omega(1). \tag{2.15}$$

When $\mathcal{I} = \mathcal{V}(g)$, the entire network is infected. In such a case, the probability of any node being the source is $1/n$.

$\square$

Back-of-Envelope Proof of Theorem 2.7.   Again assume the BFS tree rooted at the source is the regular tree defined earlier. If the depth of the tree is

$$\frac{\log n}{\log(\mu - 1)} + 1,$$

where $\mu = np$, then the number of nodes on the BFS tree is

$$1 + \sum_{j=1}^{\frac{\log n}{\log \mu - 1} + 1} \mu(\mu - 1)^{j-1} = 1 + \mu \frac{(\mu - 1)^{\frac{\log n}{\log \mu - 1} + 1} - 1}{\mu - 1} = \omega(n),$$

which implies that the BFS tree has a depth of at most $\frac{\log n}{\log(\mu - 1)} + 1$.

Now the process to generate the ER random graph and the process of the diffusion under the IC model can be viewed as a combined process. In this combined process, an edge exits only when the edge exists in the ER random graph and is live under the IC model. Loosely speaking, an edge (u, v) is said to be live if node v is infected by node u under the IC model. When the observation time is larger than or equal to the diameter of the coupled ER random graph, all nodes in the network are infected. In such a case, the probability of a node being the source is $1/n$ as the source was uniformly chosen. Based on the diameter result above, the diameter of the combine network is smaller than $\frac{\log n}{\log(npq-1)} + 1$ with probability one asymptotically. $\qquad\square$

Theorem 2.8 Consider $1 > \delta > 0$. Assuming $np > \frac{9}{\delta^2} \log n$ and $t = \frac{\log n}{\log(np-1)} - \omega(1)$, at least $1 - \delta$ fraction of the nodes in the BFS tree are leaf nodes with a high probability when n is sufficiently large. $\qquad\square$

Back-of-Envelope Proof of Theorem 2.8. Suppose the BFS from the source is the regular tree defined earlier. If the depth of the tree is $t = \frac{\log n}{\log(np)} - \omega(1)$, the size of the tree is

$$1 + \sum_{j=1}^{t} \mu(\mu-1)^{j-1} = 1 + \mu \frac{(\mu-1)^t - 1}{\mu-1} = o(n),$$

and the tree has

$$\mu(\mu-1)^{t-1}$$

leaf nodes. Therefore, the ratio of the number of the leaf nodes to the total number of nodes is

$$\frac{\mu(\mu-1)^{t-1}}{1 + \mu \frac{(\mu-1)^t - 1}{\mu-1}} \approx 1$$

for sufficiently large μ. We also remark that this argument, while not precise, is the key intuition of the formal proof when the size of the BFS tree is $o(n)$. $\qquad\square$

The theorem above suggests that the ratio between the number of leaf nodes and the total number of nodes is close to 1 so the BFS tree from the source has large BND. While the theorem does not provide any guarantee on the detection rate, it justifies the tie-breaking using BND and WBND.

2.3   JORDAN INFECTION CENTER: THE UNIVERSAL SAMPLE-PATH-BASED SOURCE ESTIMATOR

In the previous sections, we discussed source localization under the IC model and showed that the ML estimator on tree networks is a Jordan infection center and the source is the unique Jordan infection center on the ER random graph under proper conditions. It turns out that the Jordan infection center is a good source estimator for a broad range of contact-based diffusion models besides the IC model. Specifically, it has been shown in Luo et al. [2017], Zhu and Ying [2016b] that the Jordan infection center is a universal *sample-path-based* source estimator, a concept first proposed in Zhu and Ying [2013a].

We next introduce the sample-path-based approach proposed in Zhu and Ying [2013a] for the following diffusion and information models.

- **The SIR model:** Each node in the network has three possible states: susceptible (S), infected (I), and recovered (R). Nodes may change their states at the beginning of each time slot. Initially, all nodes are in state S except the source. At the beginning of each time slot, each infected node infects each of its susceptible neighbors with probability q, independent of other nodes. Each infected node recovers with probability p_r, in other words, its state changes from state I to state R with probability p_r. In addition, we assume a recovered node cannot be infected again.

- **Information model:** We assume that a complete snapshot of the network, without time-stamp information, is given. We further assume both susceptible nodes and recovered nodes are observed as healthy nodes and are indistinguishable from the observation.

"Sample path" is a standard concept used in random processes, which refers to a specific realization of the random process. Note that the diffusion process in networks is a random process. A specific diffusion trace, i.e., a sample path, is a sequence of infection and recovery timestamps that unique determine the diffusion. In this section, we focus on tree networks because the result that the Jordan infection center is a sample-path-based source estimator under a number of diffusion models were proved only for tree networks.

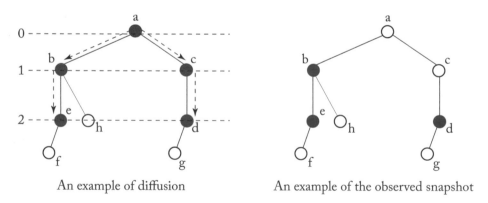

An example of diffusion   An example of the observed snapshot

Figure 2.13: Examples of information diffusion and the observed snapshot.

Consider a simple example shown in Figure 2.13. A possible sample path associated with the network for this example is

$$(a, 0, 2) \begin{array}{l} \rightarrow (b, 2, \infty) \rightarrow (e, 3, \infty) \\ \rightarrow (c, 1, 2) \rightarrow (d, 2, \infty) \end{array},$$

where the first number is the observation time, the second number is the recovery time, ∞ indicates the node has not recovered when the snapshot is observed, and "$\rightarrow$" indicates the direction of infection. Since the actual sample path is unknown, we need to consider all possible sample paths that can lead to the observed snapshot. For example, another possible sample path for Figure 2.13 is

$$(d, 0, \infty) \rightarrow (c, 1, 2) \rightarrow (a, 2, 3) \rightarrow (b, 3, \infty) \rightarrow (e, 4, \infty)$$

where the snapshot is taken right after time slot 4.

Let $\mathbf{X}[0, t]$ denote a sample path from time 0 to t, which is an $n \times (t + 1)$ matrix such that the $(\tau + 1)$th column $\mathbf{X}[\tau]$ is a vector representing the states of all nodes at time t. Let $\mathbf{Y}$ denote the observed snapshot in which each node has only two states: healthy or infected. The sample-path-based approach is to identify the *most likely* sample path $\mathbf{X}^*[0, t^*]$ that can lead to the observed snapshot, in other words, that solves the following problem:

$$\mathbf{X}^*[0, t^*] \in \arg \max_{t, \mathbf{X}[0,t]:\mathbf{F}(\mathbf{X}(t))=\mathbf{Y}} \Pr\left(\mathbf{X}[0, t]\right), \qquad (2.16)$$

where $\mathbf{F}(\cdot)$ is a mapping from the infected state to the infected state and the susceptible/recovered states to the healthy state. The source associated with $\mathbf{X}^*[0, t^*]$ is then regarded as the diffusion source.

Due to the large number of possible sample paths, solving (2.16) is computationally difficult. Zhu and Ying [2013a] established important structure properties of the optimal sample path when the network is a tree and showed that the source node of the optimal sample path must be a Jordan infection center. This property makes the problem of finding the optimal sample-path-based estimator easy. In the example, node a is a Jordan infection center because it can reach all infected nodes within two hops and it takes at least three hops for other nodes to reach all infected nodes. The complete proof has been presented in Zhu and Ying [2016b]. The key idea is the neighboring nodes lemma similar to Lemma 2.1: Consider two neighboring nodes, say node a and node b, and assume node a has a smaller infection eccentricity. Then, the optimal sample path starting from a occurs with a higher probability than the one starting from b. The proof is to construct a sample path starting from node a based on the optimal sample path starting from node b but having a higher probability to occur. After establishing the neighboring-nodes lemma, as we have shown in Lemma 2.3, on tree networks, the infection eccentricity strictly decreases along the path from any node, which is not a Jordan infection center, to a Jordan infection center. Therefore, the probabilities of the optimal sample paths associated with the nodes on the path increase monotonically by repeatedly applying the neighboring-nodes lemma.

Before presenting the neighboring-nodes lemma, we first present the following lemma without proof. The lemma says that the duration of the optimal sample path starting from node a is equal to the infection eccentricity of node a.

Lemma 2.9 *Consider a tree network with infinitely many levels. Denote by $\mathbf{Y}$ the observation we have, which contains at least one infected node. Consider node a. If $e(a, \mathcal{I}) \leq t_1 < t_2$, then the follow-*

ing inequality holds

$$\max_{\mathbf{X}_a[0,t_1]\in\mathcal{X}(t_1)} \Pr(\mathbf{X}[0,t_1]) > \max_{\mathbf{X}_a[0,t_2]\in\mathcal{X}(t_2)} \Pr(\mathbf{X}[0,t_2]),$$

where $\mathcal{X}(t) = \{\mathbf{X}[0,t]|\mathbf{F}(\mathbf{X}(t)) = \mathbf{Y}\}$ and $\mathbf{X}[0,t]$ is a diffusion sample path starting from node a. Furthermore, the infection duration of the optimal sample path starting from node a is equal to the infection eccentricity of node a, i.e.,

$$t_a^* = e(a,\mathcal{I}).$$

$\square$

We omit the proof, which can be found in Zhu and Ying [2016b]. The intuition is that a sample path with longer time duration involves more probabilistic events (infection, recovery, etc). For example, an observed infected node contributes a factor of $1 - p$ in each time slot to the overall probability of the sample path to remain infected. Therefore, a sample path over a longer duration is likely to be associated with a smaller probability. Furthermore, the minimum number of time slots for node a to reach all observed infected nodes is $e(a,\mathcal{I})$, which therefore must be the duration of the optimal sample path from node a.

Lemma 2.10   Neighboring-Nodes Lemma   *Consider a tree network with infinitely many levels. Assume the observed snapshot is $\mathbf{Y}$, which includes at least one infected node. For neighboring nodes a and b such that $e(a,\mathcal{I}) = t_a^* < t_b^* = e(b,\mathcal{I})$,*

$$\Pr(\mathbf{X}_a^*[0,t_a^*]) < \Pr(\mathbf{X}_b^*[0,t_b^*]),$$

where $\mathbf{X}_a^*[0,t_a^*]$ is the optimal sample path starting from node a, and $\mathbf{X}_b^*[0,t_b^*]$ is similarly defined. $\square$

Proof. While this lemma is similar to Lemma 2.1 in spirit, the proof is different because the live-edge-graph approach is not applicable for the SIR model. The key idea is to construct a sample path starting from a, which occurs with a higher probability than the optimal sample path starting from b. It is not difficult to see that $t_b^* = t_a^* + 1$ based on the previous lemma and the fact that a and b are neighbors.

We divide the tree into two subtrees $\mathcal{T}_a^{-b}$ and $\mathcal{T}_b^{-a}$ by partitioning the tree along the edge (a,b). See an example in Figure 2.14. The diffusion processes on these two subtrees are mutually independent after both a and b become infected because any path connecting a node in $\mathcal{T}_a^{-b}$ to a node in $\mathcal{T}_b^{-a}$ has to traverse edge (a,b).

To prove this lemma, we start from the optimal sample path rooted at b, $\mathbf{X}_b^*[0,t_b^*]$ and construct a sample path starting from a, denoted by $\mathbf{X}_a[0,t]$, which occurs with a higher probability. Consider the example in Figure 2.14 and assume $p_r = q = 0.5$. Then the optimal sample

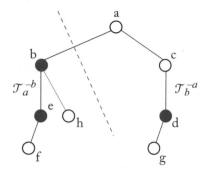

Figure 2.14: Example for illustrating the proof.

path rooted at b^*, $\mathbf{X}_b^*[0, t_b^*]$, has duration $t_b^* = 3$, and is

$$(b, 0, \infty) \quad
\begin{aligned}
&\to (e, 1, \infty) \to (f, 2, 3)\\
&\to (h, 1, 2)\\
&\to (a, 1, 2) \to (c, 2, 3) \to (d, 3, \infty)
\end{aligned}.$$

Table 2.1 summarizes the states of the nodes in this sample path over three time slots.

Table 2.1: The node states in sample path $\mathbf{X}_b^*[0, t_b^*]$

	b	a	c	d	g	e	h	f
0	I	S	S	S	S	S	S	S
1	I	I	S	S	S	I	I	S
2	I	R	I	S	S	I	R	I
3	I	R	R	I	S	I	R	R

Given $\mathbf{X}_b^*[0, t_b^*]$, we define $\mathbf{X}_b^*[0, t_b^*] \bigcap \mathcal{T}_a^{-b}$ to be the sample path restricted on subtree $\mathcal{T}_a^{-b}$, and $\mathbf{X}_b^*[0, t_b^*] \bigcap \mathcal{T}_b^{-a}$ is similarly defined. In the example above, we have

$$\mathbf{X}_b^*[0, t_b^*] \bigcap \mathcal{T}_b^{-a} : (b, 0, \infty) \quad
\begin{aligned}
&\to (e, 1, \infty) \to (f, 2, 3)\\
&\to (h, 1, 2)
\end{aligned},$$

and

$$\mathbf{X}_b^*[0, t_b^*] \bigcap \mathcal{T}_a^{-b} : (a, 1, 2) \to (c, 2, 3) \to (d, 3, \infty).$$

Now we proceed to construct a sample path starting from node a with duration of $\tilde{t}_a = t_b^* - 1 = 2$. Note that in $\mathbf{X}_b^*[0, t_b^*]$, it takes at least one time slot for node h to infect node a, so the duration of $\mathbf{X}_b^*[0, t_b^*] \cap \mathcal{T}_a^{-b}$ is at most $t_b^* - 1$. Furthermore, the infection eccentricity of node a is $e(a, \mathcal{I}) = e(b, \mathcal{I}) - 1 = t_b^* - 1$, so $\mathbf{X}_b^*[0, t_b^*] \cap \mathcal{T}_a^{-b}$ takes at least $t_b^* - 1$ time slots. Therefore, $\mathbf{X}_b^*[0, t_b^*] \cap \mathcal{T}_a^{-b}$ takes precisely $t_b^* - 1 = e(a, \mathcal{I})$ time slots. Therefore, we construct $\mathbf{X}_a[0, t_b^* - 1] \cap \mathcal{T}_a^{-b}$ such that $\mathbf{X}_a[0, t_b^* - 1] \cap \mathcal{T}_a^{-b}$ is identical to $\mathbf{X}_b^*[0, t_b^*] \cap \mathcal{T}_a^{-b}$ except that all events in $\mathbf{X}_a[0, t_b^* - 1] \cap \mathcal{T}_a^{-b}$ occur one time slot earlier than $\mathbf{X}_b^*[0, t_b^*] \cap \mathcal{T}_a^{-b}$. For the example shown in Figure 2.14, we obtain

$$\mathbf{X}_a[0, t_b^* - 1] \cap \mathcal{T}_a^{-b} : (a, 0, 1) \to (c, 1, 2) \to (d, 2, \infty).$$

Since node a needs to go through node b to reach infected nodes on $\mathcal{T}_b^{-a}$, therefore,

$$d(b, \mathcal{I} \cap \mathcal{T}_b^{-a}) \le e(a, \mathcal{I}) - 1.$$

Therefore, when restricted to subtree $\mathcal{T}_b^{-a}$, according to Lemma 2.9, there exists a sample path $\tilde{\mathbf{X}}_b[0, e(a, \mathcal{I}) - 1]$ such that

$$\Pr\left(\tilde{\mathbf{X}}_b[0, e(a, \mathcal{I}) - 1] \cap \mathcal{T}_b^{-a}\right) > \Pr\left(\mathbf{X}_b^*[0, t_b^*] \cap \mathcal{T}_b^{-a}\right).$$

For the example in Figure 2.14, we have

$$\tilde{\mathbf{X}}_b[0, 1] \cap \mathcal{T}_b^{-a} : (b, 0, \infty) \quad \begin{array}{l} \to (e, 1, \infty) \\ \to (h, \infty, \infty) \end{array}.$$

Therefore, we construct $\mathbf{X}_a[0, t_b^* - 1] \cap \mathcal{T}_b^{-a}$ such that it is identical to $\tilde{\mathbf{X}}_b[0, e(a, \mathcal{I}) - 1]$ except all events in $\mathbf{X}_a[0, t_b^* - 1] \cap \mathcal{T}_b^{-a}$ occur one time slot later. For the example in Figure 2.14, we have

$$\mathbf{X}_a[0, t_b^* - 1] \cap \mathcal{T}_b^{-a} : (b, 1, \infty) \quad \begin{array}{l} \to (e, 2, \infty) \\ \to (h, \infty, \infty) \end{array}.$$

In summary, we construct a sample path from node a such that

$$
\begin{aligned}
\Pr\left(\mathbf{X}_a[0, t_a^*]\right) &= q \Pr\left(\tilde{\mathbf{X}}_b[0, e(a, \mathcal{I}) - 1] \cap \mathcal{T}_b^{-a}\right) \Pr\left(\mathbf{X}_b^*[0, t_b^*] \cap \mathcal{T}_a^{-b}\right) \\
&\ge q \Pr\left(\mathbf{X}_b^*[0, t_b^*] \cap \mathcal{T}_b^{-a}\right) \Pr\left(\mathbf{X}_b^*[0, t_b^*] \cap \mathcal{T}_a^{-b}\right) \\
&= \Pr\left(\mathbf{X}_b^*[0, t_b^*]\right).
\end{aligned}
$$

So the lemma holds.

For the example in Figure 2.14, the sample path from node a we can construct is

$$(a, 0, 1) \quad \begin{array}{l} \to (b, 1, \infty) \quad \begin{array}{l} \to (e, 2, \infty) \\ \to (h, \infty, \infty) \end{array} \\ \to (c, 1, 2) \to (d, 2, \infty) \end{array}.$$

$\square$

Theorem 2.11   Assume the SIR model for diffusion and infinite tree networks. The root of the optimal sample path that solves problem (2.16) must be a Jordan infection center and the duration of the optimal sample path is equal to the infection eccentricity of the Jordan infection center.

Proof. Based on Lemma 2.10, we have that for two neighboring nodes, say node a and node b, if node b has a larger infection eccentricity than that of node a, then node b is not the root of the optimal sample path. Now recall Lemma 2.3, which states that on tree networks, the infection eccentricity decreases from any node to the Jordan infection center closer to the node. Therefore, we conclude that the Jordan infection center is the root of the mostly likely sample path. The duration of the optimal sample path is equal to the infection eccentricity of the Jordan infection center according to Lemma 2.9.   □

Besides being the sample-path-based estimator, the Jordan infection center also has the following performance guarantee.

Theorem 2.12   Consider a $(g + 1)$-regular tree with infinitely many levels. Assume $g > 2$, $gq > 1$, and the observed snapshot $\mathbf{Y}$ includes at least one infected node. Given $\epsilon > 0$, there exists a constant d_ϵ such that the distance between the Jordan infection center and the actual source is d_ϵ with probability $1 - \epsilon$, where d_ϵ is independent of the number of infected nodes and the time at which the snapshot $\mathbf{Y}$ was taken.

The theorem states that the distance between the Jordan infection center and the actual source is $O(1)$ where the scaling is with respect to the number of infected nodes. This is a desired property for large-scale networks because it reduces the search space to a small neighborhood of the Jordan infection center. The proof can be found in Zhu and Ying [2016b]. The idea is to show that with a high probability, one of the following events will occur: (1) the infection process terminates within d hops from the source; (2) at least n_0 nodes are infected in the d-hop neighborhood of the source. When the first event occurs, it is easy to see that a Jordan infection center is within a constant distance from the actual source. When the second event occurs, we consider the infection processes starting from the infected nodes that are within d-hop away from the source. For each node, we consider a branching process in which a node's offsprings are the nodes that are immediately infected after the node is infected. So the branching process grows one level at each time slot. We show that with a high probability, two branching processes starting from the nodes within d-hop from the source survive, where "survive" means the branching process has not terminated when the observation was taken. The existence of two survived branching processes guarantees that we have at least two observed infected nodes that are $\approx 2t - O(1)$ apart from each other, which guarantees that the minimum infection eccentricity of the network is $t - O(1)$. Since the infection eccentricity of the source node is at most t, we can then conclude that a Jordan infection center, which has an infection eccentricity $t - O(1)$, is not far from the actual source by using the fact the path between two nodes is unique on a tree.

The results presented above are for the SIR model with a complete snapshot and have been extended to several directions, which are summarized below.

- **Other diffusion models:** Luo et al. [2017] proved that on infinite tree networks, Jordan infection center is also the optimal sample-path-based estimator for the Susceptible-Infected-Recovered-Infected (SIRI) and Susceptible-Infected-Susceptible (SIS) models. This result, combined with the results presented in this section, show that the Jordan infection center is a sample-path-based estimator that does not depend on the infection, recovery and reinfection rates, so is a robust source estimator.

- **Sparse observation**: Zhu and Ying [2013b] considered source localization with sparse observations under the SIR model and assumed that a small subset of infected nodes are reported. For infinite trees, it is proved that the sample-path-based estimator is a Jordan infection center with respect to the set of observed infected nodes. Furthermore, the distance between the estimator and the actual source is upper bounded by a constant independent of the number of infected nodes with a high probability.

2.4   NETSLEUTH: APPROXIMATE ML ESTIMATOR FOR GENERAL NETWORKS

In this section, we present another interesting approach for locating diffusion sources called Netsleuth [Prakash et al., 2012]. The approach is based on mean field approximation to solve the ML problem for general networks. We consider the discrete time SI (susceptible-infected) model and a complete snapshot of the network, where the SI model is a special case of the SIR model where infected nodes do not recover.

2.4.1   MEAN FIELD APPROXIMATION

Recall the ML estimator for the source is the solution of the following problem:

$$\arg\max_{v \in \mathcal{I}} \Pr(\mathcal{O}|v).$$

The computational complexity of solving the ML estimator is prohibitively high for general networks. Therefore, mean field approximation has been used to find a heuristic solution. The approximation calculates the marginal probability of a node being infected and assumes a node is infected independent of other nodes' states.

In particular, define $X_u(t)$ to be the state of node u at time t where $X_u(t) = 1$ means the node is infected and 0 means the node is susceptible. The mean field approximation assumes that

$$\Pr(\mathcal{O}|v) \approx \prod_{u \in \mathcal{I}} \Pr(X_u = 1|v). \tag{2.17}$$

Next, we focus on computing the marginal probability that node u is infected when node v is the source based on the mean field approximation. It will be shown that the probability can be approximated using the eigenvector associated with the largest eigenvalue of a matrix defined by the graph and infection subgraph.

We now introduce the following notation.

- Define $Y_{uv}(t)$ to be the indicator whether node u can successfully infect node v at time t. We assume $Y_{uv}(t)$ are independent across edges and time.

- Define $\mathbf{D}$ to be the degree matrix of graph g, i.e.,

$$\mathbf{D} = \begin{bmatrix} d_1 & 0 & \dots & \dots & 0 \\ 0 & d_2 & 0 & \dots & 0 \\ & & \dots\dots\dots & & \\ 0 & 0 & \dots & 0 & d_n \end{bmatrix}$$

where d_u is the degree of node u.

- Define $\mathbf{A}$ to be the adjacency matrix of graph g such that $A_{uv} = 1$ if nodes u and v are connected.

- The Laplacian matrix of graph g is

$$\mathbf{L} = \mathbf{D} - \mathbf{A}.$$

- Denote by $\mathbf{L}_{\mathcal{I}}$ the sub matrix of $\mathbf{L}$ that includes only infected nodes.

Consider node u. Node u is in the infected state at time $t + 1$ if either node u was infected before time $t + 1$ or node u was infected at time $t + 1$ by one of its infected neighbors. Therefore, we have

$$\Pr\left(X_u(t+1) = 1\right)$$

$$= \Pr\left(X_u(t) = 1\right) + \Pr\left(X_u(t) = 0\right) \Pr\left(\bigcup_{v:(v,u)\in\mathcal{E}} \{X_v(t)Y_{vu}(t+1) = 1\}\right),$$

where $\bigcup$ is the "OR" operation. In the equation above, $X_v(t)Y_{vu}(t + 1) = 1$ if $X_v(t) = 1$, i.e., node v is in the infected state at time slot t, and $Y_{vu}(t + 1) = 1$, i.e., node v successfully infected node u at time slot $t + 1$. Therefore, $\bigcup_{v:(v,u)\in\mathcal{E}}\{X_v(t)Y_{vu}(t + 1) = 1\}$ if node u is infected by any of its neighbors at time slot $t + 1$, which implies

$$\Pr(X_u(t + 1) = 1)$$

$$\leq \Pr(X_u(t) = 1) + \Pr\left(\bigcup_{v:(v,u)\in\mathcal{E}}\{X_v(t)Y_{vu}(t + 1) = 1\}\right)$$

$$\leq_{(a)} \Pr(X_u(t) = 1) + \sum_{v:(v,u)\in\mathcal{E}} \Pr\left(X_v(t)Y_{vu}(t + 1) = 1\right)$$

$$= \Pr(X_u(t) = 1) + \sum_{v:(v,u)\in\mathcal{E}} \Pr\left(X_v(t) = 1\right)\Pr\left(Y_{vu}(t + 1) = 1|X_v(t) = 1\right),$$

where inequality (a) follows from the union bound, and the last equality holds because $X_v(t)$ and $Y_{vu}(t)$ are independent.

Note that $\Pr\left(Y_{uv}(t + 1)|X_v(t) = 1\right) \leq A_{uv}$ because node v cannot infect node u if the two nodes are not neighbors. Therefore, we have

$$\sum_{v:(v,u)\in\mathcal{E}} \Pr\left(X_v(t) = 1\right)\Pr\left(Y_{vu}(t + 1) = 1|X_v(t) = 1\right)$$

$$\leq \sum_{v:(v,u)\in\mathcal{E}} A_{uv}\Pr(X_v(t) = 1)$$

$$\leq \sum_{v:(v,u)\in\mathcal{E}} A_{uv}\left(\Pr(X_v(t) = 1) - \Pr(X_u(t) = 1) + \Pr(X_u(t) = 1)\right)$$

$$= \sum_{v:(v,u)\in\mathcal{E}} A_{uv}\left(\Pr(X_v(t) = 1) - \Pr(X_u(t) = 1)\right) + \sum_{v:(v,u)\in\mathcal{E}} A_{uv}\Pr(X_u(t) = 1).$$

Letting $d_{\max}$ be the maximum degree of the graph, we have

$$\sum_{v:(v,u)\in\mathcal{E}} A_{uv}\Pr(X_u(t) = 1) \leq d_{\max}\Pr(X_u(t) = 1).$$

Based on the equations above, we have

$$\Pr(X_u(t + 1) = 1) \tag{2.18}$$

$$\leq \Pr\left(X_u(t) = 1\right) + d_{\max}\Pr(X_u(t) = 1) + \sum_{v:(v,u)\in\mathcal{E}} A_{uv}\left(\Pr(X_v(t) = 1) - \Pr(X_u(t) = 1)\right)$$

$$\tag{2.19}$$

$$= \sigma\Pr(X_u(t) = 1) + \sum_{v:(v,u)\in\mathcal{E}} A_{uv}\left(\Pr(X_v(t) = 1) - \Pr(X_u(t) = 1)\right), \tag{2.20}$$

where $\sigma = 1 + d_{\max}$.

Define $\mathbf{P}(t) = (\Pr(X_1(t) = 1), \cdots, \Pr(X_n(t) = 1))$ and

$$\mathbf{M} \triangleq \begin{bmatrix} \mathbf{0}_{n-|\mathcal{I}|,n-|\mathcal{I}|} & \mathbf{0}_{n-|\mathcal{I}|,|\mathcal{I}|} \\ \mathbf{0}_{|\mathcal{I}|,n-|\mathcal{I}|} & \mathbf{L}_{\mathcal{I}} \end{bmatrix}$$

where $\mathbf{0}_{m,n}$ is a zero matrix with size $m \times n$. We reorganize Equation (2.20) into a matrix form:

$$\mathbf{P}(t+1) \leq \sigma \left(\mathbf{I} - \frac{1}{\sigma} \mathbf{M} \right) \mathbf{P}(t), \tag{2.21}$$

where $\mathbf{I}$ is an identity matrix. Denote by $\mathbf{P}_{\mathcal{I}}(t)$ the sub vector of $\mathbf{P}(t)$ which only includes infected nodes. We have

$$\mathbf{P}_{\mathcal{I}}(t+1) \leq \sigma \left(\mathbf{I} - \frac{1}{\sigma} \mathbf{L}_{\mathcal{I}} \right) \mathbf{P}_{\mathcal{I}}(t) \tag{2.22}$$

$$\leq \sigma^t \left(\mathbf{I} - \frac{1}{\sigma} \mathbf{L}_{\mathcal{I}} \right)^t \mathbf{P}_{\mathcal{I}}(0) \tag{2.23}$$

$$= \sigma^t \sum_i \lambda_i^t \mathbf{x}_i \mathbf{x}_i' \mathbf{P}_{\mathcal{I}}(0), \tag{2.24}$$

where λ_i and $\mathbf{x}_i$ the ith eigenvalue and eigenvector of matrix $\mathbf{I} - \frac{1}{\sigma} \mathbf{L}_{\mathcal{I}}$ and $\mathbf{x}'$ is the transpose of $\mathbf{x}$.

From (2.24), we can see when t is large, the upper bound is dominated by the term associated with the largest eigenvalue, which can be used to estimate $\mathbf{P}_{\mathcal{I}}(t+1)$. The next lemma shows that the largest eigenvalue and eigenvector are all positive.

Lemma 2.13 *The largest eigenvalue λ_1 and the associated eigenvector $\mathbf{x}_1$ of the matrix $\mathbf{I} - \frac{1}{\sigma} \mathbf{L}_{\mathcal{I}}$ are positive and real.*

Proof. Matrix $\mathbf{I} - \frac{1}{\sigma} \mathbf{L}_{\mathcal{I}}$ is non-negative. Since $g_{\mathcal{I}}$ is connected under the SI model, if we view $\mathbf{I} - \frac{1}{\sigma} \mathbf{L}_{\mathcal{I}}$ as an adjacency matrix of a graph, then the graph is irreducible. Based on the Perron-Frobenius theorem [MacCluer, 2000], the lemma holds. □

Based on Lemma 2.13, we can then conclude that

$$\mathbf{P}_{\mathcal{I}}(t+1) \leq \sigma^t \lambda_1^t \sum_i \frac{\lambda_i^t}{\lambda_1^t} \mathbf{x}_i \mathbf{x}_i' \mathbf{P}_{\mathcal{I}}(0) \tag{2.25}$$

$$\approx \sigma^t \lambda_1^t \mathbf{x}_1 \mathbf{x}_1' \mathbf{P}_{\mathcal{I}}(0) \tag{2.26}$$

for sufficiently large t when $\lambda_1 > \lambda_2$. For single-source diffusion, we know that $\mathbf{P}_{\mathcal{I}}(0)$ is all zero except for the source. We conclude with the following approximation:

$$\Pr(X_u = 1|s) \propto x_{1,u} x_{1,s}, \forall u \in \mathcal{I}, \tag{2.27}$$

$$\Pr(X_u = 1|s) = 0, \forall u \notin \mathcal{I}, \tag{2.28}$$

where $x_{1,u}$ is the uth entry of eigenvector $\mathbf{x}_1$.

Recall that under the mean field approximation, we assume

$$\Pr(\mathcal{O}|s) \approx \prod_{u \in \mathcal{I}} \Pr(X_u = 1|s), \tag{2.29}$$

which implies

$$\log \Pr(\mathcal{O}|s) \approx \sum_{u \in \mathcal{I}} \log \Pr(X_u = 1|s) \tag{2.30}$$

$$\propto \sum_{u \in \mathcal{I}} \log x_{1,u} + |\mathcal{I}| \log x_{1,s} \tag{2.31}$$

$$\propto \log x_{1,s} \tag{2.32}$$

$$\propto x_{1,s}. \tag{2.33}$$

The result above suggests that the probability of observing $\mathcal{O}$ given s as the source is proportional to the value of $x_{1,s}$, which is the sth entry of the largest eigenvector of matrix $\mathbf{I} - \frac{1}{\sigma}\mathbf{L}_{\mathcal{I}}$. Motivated by the discussion above, Prakash et al. [2012] suggests the following ML estimator

$$s^* \in \arg\max_u x_{1,u},$$

and the following algorithm, named NETSLEUTH.

Algorithm 2.3 The Single-Source NetSleuth Algorithm

1: Compute the degree matrix $\mathbf{D}$.
2: Compute the Laplacian matrix $\mathbf{L} = \mathbf{D} - \mathbf{A}$.
3: Extract submatrix $\mathbf{L}_{\mathcal{I}}$ based on $\mathcal{I}$.
4: Compute the largest eigenvalue and the corresponding eigenvector $\mathbf{x}_1$.
5: **return** The node s^* such that $s^* \in \arg\max_u x_{1,u}$.

2.5   NUMERICAL EVALUATION

In this section, we compare the performance of wSFT, SFT, Jordan, RUM (rumor centrality), and NETSLEUTH, and on different networks such as tree networks, the ER random graph, and real world networks. We defer the detailed discussion of RUM and its performance to Chapter 3.

- **Jordan:** Select the node with minimum infection eccentricity, e.g., the Jordan infection center. Ties are broken randomly. Recall that the optimal sample path estimator on tree networks is a Jordan infection center for a number of different diffusion models.

- **RUM:** Select the node with maximum rumor centrality proposed in Shah and Zaman [2011]. The rumor centrality was proved to be the maximum likelihood estimator on regular trees under the *continuous-time* SI model in which the infection time follows exponential distribution. Shah and Zaman [2011] is also the first paper that analytically studied the source localization problem.

Among the selected algorithms, only wSFT requires the infection probabilities. We included wSFT to evaluate the importance of the knowledge of edge weights for source localization. We will see that the performance of SFT is almost identical to wSFT. Note that these numerical results were originally presented in Zhu and Ying [2015].

2.5.1 EVALUATION METRICS

We evaluated the performance of the algorithms with three different metrics.

- **Detection Rate:** Detection rate is the probability that the node identified by the algorithm is the actual source. A desired goal of diffusion source localization is to have a high detection rate.

- **Distance to the Source:** Distance is the number of hops from the source estimator to the actual source. The distance provides the accuracy of the estimator even when the estimator does not coincide with the source.

- $\gamma\%$-**Accuracy:** $\gamma\%$-accuracy is the probability with which the source is ranked among the top γ percent. Note that besides providing a source estimator, a source localization algorithm can also be used to rank the infected nodes according to their likelihood to be the source. For example, SFT can rank the nodes in ascending order according to their infection eccentricity and then breaks ties using BND. Other algorithms can be used to rank nodes as well. $\gamma\%$-accuracy is a less ambitious alternative to the detection rate. When the detection rates of all algorithms are low, it is useful to compare $\gamma\%$-accuracy because a high $\gamma\%$-accuracy guarantees that the actual source is among the top ranked nodes with a high probability.

2.5.2 BINOMIAL TREES

In this section, we evaluate the algorithms on binomial trees. Denote by $Bi(m, \beta)$ the binomial distribution with m number of trials, and each trial succeeds with probability β. A binomial tree is a tree where the number of children of each node follows a binomial distribution $Bi(m, \beta)$. In the experiments, we set $m = 20$ and $\beta = 0.5$. We adopted the IC model where the infection probability of each edge is assigned with a uniform distribution in $(0.2, 0.5)$. The lower bound on the infection probability is set to be 0.2 to prevent the diffusion process from dying out quickly. We evaluated the performance for different infection size x. Under a discrete infection model, it is hard to obtain the diffusion snapshots with exact x infected nodes. Therefore, for each infection size x, we generate the diffusion samples where the number of infected nodes are in range $[0.75x, 1.25x]$. The source was chosen uniformly at random among all nodes in the network. We varied x from 200 to 2,000 with a step size 200. For each infection size, we generated 400 diffusion samples.

Figure 2.15 shows the detection rates for different infection sizes. The detection rates of Jordan, SFT, and wSFT do not change for different infection sizes since the structure of the binomial tree is simple. SFT, wSFT, and Jordan have the highest detection rate (more than 0.9), while the detection rates of RUM and NETSLEUTH are much lower. The distance results are shown in Figure 2.16. As expected, SFT, wSFT, and Jordan outperform RUM, which are all much better than NETSLEUTH. Figure 2.17 shows the $\gamma\%$-accuracy vs. the rank percentage γ. We picked infection size 1,000. As shown in Figure 2.17, all three algorithms based on infection eccentricity (Jordan, SFT, wSFT) have better performance than RUM and NETSLEUTH. Recall that the node identified by wSFT is an ML estimator of the actual source.

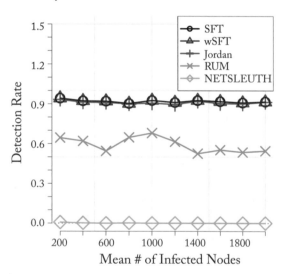

Figure 2.15: Detection rates of different source localization algorithms on the binomial trees.

2.5.3   THE ER RANDOM GRAPH

In this section, we compared the performance of the algorithms on the ER random graph. In the experiments, we generated the ER random graph with $n = 5,000$ and wiring probability $p = 0.002$. We again varied the infection network size from 200 to 2,000. The infection probability of each edge is assigned with a uniform distribution in $(0.2, 0.5)$. We generated 400 diffusion samples.

Figure 2.18 shows the detection rate vs. the infection size. The detection rate decreases as the infection size increases. SFT and wSFT have higher detection rates compared to other

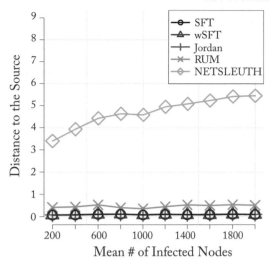

Figure 2.16: Distance to the source of different source localization algorithms on the binomial trees.

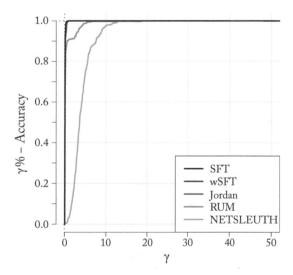

Figure 2.17: γ%-accuracy of different source localization algorithms on the binomial trees.

algorithms. Figure 2.19 shows the results on distance. As we expected, SFT and wSFT out-perform other algorithms when the infection size is less than 1,600 nodes. As the size of the infected nodes increases, SFT and wSFT become close to RUM in term of distance to the source. However, the detection rate of both algorithms are still much higher than that of RUM.

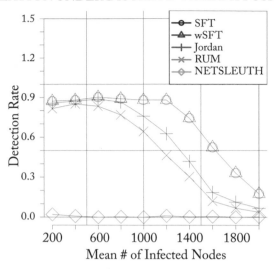

Figure 2.18: Detection rates of different source localization algorithms in the ER random graph.

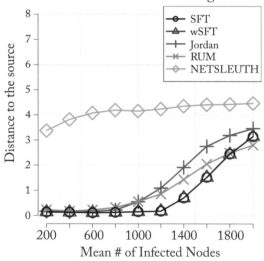

Figure 2.19: Distance to the source of different source localization algorithms in the ER random graph.

Another observation is that SFT and wSFT have identical performance, which indicates that the performance of SFT is robust to edge weights.

Figure 2.20 shows the $\gamma\%$-accuracy vs. the rank percentage γ with 1,000 infected nodes. SFT and wSFT have similar or better performance compared to all other algorithms.

Although the performance of Jordan and SFT algorithms are similar on tree networks, SFT outperforms Jordan significantly on the ER random graph. The observation indicates that BND is an effective tie-breaking rule and increases the detection accuracy.

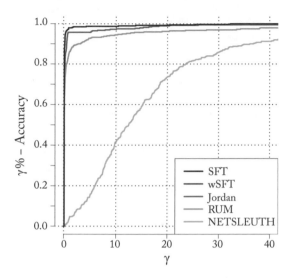

Figure 2.20: γ%-accuracy of different source localization algorithms in the ER random graph.

2.5.4   THE INTERNET AUTONOMOUS SYSTEM NETWORK

The Internet autonomous systems (IAS) network[1] is the Internet autonomous system from Oregon route views on March 31, 2001 with 10,670 nodes and 22,002 edges. The IAS network is a small world network. We adopted similar settings as in Section 2.5.3.

The detection rates are shown in Figure 2.21. The detection rate of Jordan is low since the IAS graph is a small world network and there are multiple Jordan infection centers due to the small diameter of the network. With the tie-breaking rule BND, the detection rate doubles in most cases, which demonstrates the effectiveness of BND. While the detection rate of SFT is only 10% when the infection size is 1,000, the distance to the actual source is slightly more than one hop, as shown in Figure 2.22. In addition, the γ%-accuracy vs. γ for 1,000 infection size is shown in Figure 2.20. The 10% accuracies of SFT and wSFT are close to 70%, which are significantly higher than that of other algorithms.

2.5.5   RUNNING TIME VS. PERFORMANCE

In this section, we evaluated the scalability of the algorithms by comparing the running time. The experiments were conducted on an Intel Core i5-3210M CPU with four cores and 8G

[1]Available at http://snap.stanford.edu/data/index.html

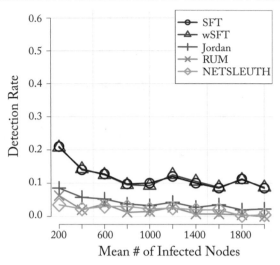

Figure 2.21: Detection rates of different source localization algorithms in the IAS graph.

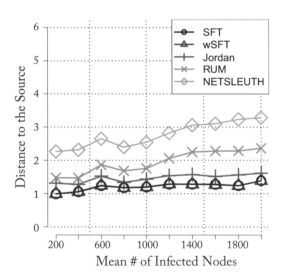

Figure 2.22: Distance to the source of different source localization algorithms in the IAS graph.

RAM with a Windows 7 Professional 64-bit system. All algorithms were implemented with Python 2.7. The ER random graphs with 5,000 nodes and $p = 0.002$ edge generation probability were used in the experiments. The infection probability of each edge is uniformly distributed over $(0.2, 0.5)$. We generated 100 diffusion samples for the experiments. Figure 2.24 show the average running time vs. the detection rate. The infection size is chosen to be 1,000. SFT and wSFT

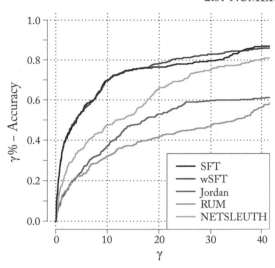

Figure 2.23: γ%-accuracy of different source localization algorithms in the IAS graph.

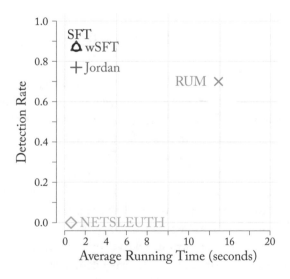

Figure 2.24: Detection rate vs. running time in the ER random graph.

took 1.11 seconds and achieves 0.87 detection rate while NETSLEUTH took 0.62 seconds with 0 detection rate and RUM took 14.86 seconds with 0.7 detection rate. The detection rate of SFT is much higher than NETSLEUTH and SFT is 14 times faster than RUM.

2.6   LOCATING MULTIPLE DIFFUSION SOURCES

So far, we assumed that the diffusion is from single source. In practice, the diffusion can be originated from multiple nodes instead of from a single source. When the diffusion duration is sufficiently short, the infected subnetworks from different sources are disconnected components. In such cases, the single-source localization algorithms can be applied to each of the infected subnetworks.

When the diffusion duration is not short, then likely the infected subnetworks merge and overlap with each other. In this case, single-source localization algorithms cannot be directly applied, but can be adopted. One straightforward approach is to combine single-source estimators with Kmeans and use iterative clustering and localization algorithms [Chen et al., 2016, Luo et al., 2013, Zhu et al., 2017]. In particular, assuming that diffusion processes from all the sources start at the same time and the number of sources is known (say m), the clustering and localization repeats the following two steps after select m nodes as the initial source estimators.

- Clustering: Use the m source estimators as "centers" and partition the network into m clusters (subnetworks) by assigning each node to one of the centers according to some distance measure.

- Localization: For each cluster, identify a source estimator using one of single-source localization algorithms.

We note that NETSLEUTH was developed for locating multiple sources [Prakash et al., 2012]. For locating multiple sources, it works as follows: First, it identifies the first source estimator as discussed in Section 2.4. Then, the state of the source estimator is changed to healthy, and the same algorithm is used again on the updated graph to obtain the second source estimator. The process repeats until m source estimators are identified.

When diffusion processes from different sources start from different times, the problem becomes more challenging as the infection subgraphs may have very different sizes, and the clustering and localization approach mentioned above does not work. A novel algorithmic framework to address this problem has been recently proposed in Ji et al. [2017] where the concepts of heavy centers and covering are introduced to overcome the issue of heterogeneous sizes of infection subgraphs from different sources. A heavy center is the infection subgraph under a deterministic diffusion from a single source, and covering differs from partition by allowing overlap in partition. The algorithm then uses heavy centers and covering as in the clustering and localization approach. Interested readers can find the details in Ji et al. [2017].

Finally, when the number of sources is unknown, algorithms for estimating the number of possible sources have been developed in Chen et al. [2016], Ji et al. [2016].

CHAPTER 3

Source Localization under Continuous-Time Diffusion Models

This chapter focuses on source localization under continuous-time diffusion models. We first present the basic models and then discuss rumor centrality, introduced in a seminal paper [Shah and Zaman, 2011].

3.1 RUMOR CENTRALITY: THE ML ESTIMATOR

3.1.1 MODELS

- Network model: We consider an undirected tree network, denoted by g.

- Diffusion model: We consider the susceptible-infected (SI) model and assume the time is continuous. Each node has two possible states: susceptible and infected. A susceptible node may be infected by an infected neighbor. Denote by τ_{ij} the amount of time it takes to spread the infection from node i to node j. We assume that τ_{ij} is an exponential random variable with mean λ and τ_{ij}'s are independent across edges. At time 0, all nodes in the network are in the susceptible state except the source.

- Information model: Similar to Chapter 2, we assume that a complete snapshot $\mathcal{O} = \{\mathcal{I}, \mathcal{H}\}$ of the network taken at time t is given.

Example
Figure 3.1 shows a diffusion example under the continuous-time SI model. The values next to the edges are the infection spreading times. In this example, node b was infected at time 0.7 by node s and node d was infected at time $0.7 + 0.2 = 0.9$ by node b. At time 0.6, node a was infected. At time 1.4, nodes c, e, b, and d were infected. Under the continuous-time SI model, all nodes will be infected eventually.

3.1.2 PROBLEM FORMULATION
Similar to the notation used in Chapter 2, denote by $\mathcal{E}(g)$ the set of edges in g and $\mathcal{V}(g)$ the set of nodes in g. Based on $\mathcal{O}$, the goal is to find the diffusion source in the network. The infected

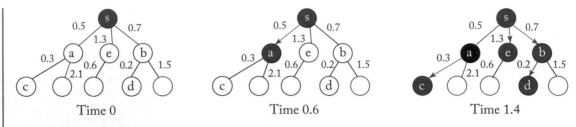

Figure 3.1: An example to illustrate the continuous-time SI model.

nodes form a connected component under the continuous-time SI model, called the infection subgraph and denoted by g_i. We assume the source is chosen uniformly at random from $\mathcal{V}(g)$. Based on the notation above, the source localization problem can be formulated as a maximum likelihood problem:

$$\arg\max_{v \in \mathcal{I}} \Pr(\mathcal{O}|v), \tag{3.1}$$

where $\Pr(\mathcal{O}|v)$ is the probability of having snapshot $\mathcal{O}$ given that v is the source.

In general networks, it is difficult to calculate $\Pr(\mathcal{O}|v)$. We next introduce an important structure property of the ML estimator on tree networks [Shah and Zaman, 2011].

Note that under the continuous-time SI model, nodes are infected at different time instances because the probability that two nodes are infected at the same time is zero. Therefore, given the infected subgraph g_i, we call a permutation of $\mathcal{V}(g_i)$ a *permitted permutation* if it is a valid infection sequence that can lead to g_i.

Consider the simple example in Figure 3.2. The red nodes are infected nodes. In this example, (a, b, c, d) is a permitted permutation and it implies that a first infected b, then b infected c, and finally c infected d. On the other hand, (a, c, b, d) is not a permitted permutation because c cannot be infected by a directly. Note that for each node, say node u, in a permitted permutation, at least one of the nodes before node u should be an infected neighbor of node u.

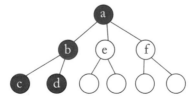

Figure 3.2: An example of a tree with four infected nodes.

Given infection subgraph g_i with size m, define $\mathbf{V} = (V_1, \cdots, V_m)$ to be the sequence in these m infected nodes were infected, which is a random vector. We next calculate the probability of $\mathbf{V} = \mathbf{v} = (v_0, v_1, \cdots, v_m)$ when $\mathbf{v}$ is a permitted permutation. Given a sequence of nodes $\mathbf{v}$, we define $\mathcal{N}(\mathbf{v})$ to be the set of nodes that are neighbors of nodes in $\mathbf{v}$, but not in $\mathbf{v}$. For example in Figure 3.2,

$$\mathcal{N}((a, b)) = \{c, d, e, f\}.$$

Due to the memoryless property of exponential distributions, given that the set of nodes currently infected is $\mathcal{B}$, nodes in $\mathcal{N}(\mathcal{B})$ are equally likely to be the next infected node. In the example in Figure 3.2,

$$\mathrm{Pr}\,(V_3 = c | (V_1 = a, V_2 = b)) = \frac{1}{4}.$$

Therefore, we have

$$\mathrm{Pr}\,(\mathbf{V} = \mathbf{v}) = \mathrm{Pr}\,(V_1 = v_1) \prod_{j=2}^{m} \mathrm{Pr}\,\left(V_j = v_j | (V_1 = v_1, \cdots, V_{j-1} = v_{j-1})\right)$$

$$= \frac{1}{|\mathcal{V}(g)|} \prod_{j=2}^{m} \frac{1}{\left|\mathcal{N}((v_1, \cdots, v_{j-1}))\right|}. \tag{3.2}$$

Consider the example in Figure 3.2,

$$\mathrm{Pr}\,(\mathbf{V} = (a, b, c, d)) = \frac{1}{|\mathcal{V}(g)|} \frac{1}{\mathcal{N}((a))} \frac{1}{\mathcal{N}((a, b))} \frac{1}{\mathcal{N}((a, b, c))} = \frac{1}{10} \times \frac{1}{3} \times \frac{1}{4} \times \frac{1}{3}.$$

From (3.2), we have that $\mathrm{Pr}\,\left(V_j = v_j | (v_1, \cdots, v_{j-1})\right)$ depends on the size of $\mathcal{N}((v_1, \cdots, v_{j-1}))$. Note that after adding a new infected node, say v_j, to the infection sequence, the number of nodes that can be infected at next instance increases by $d_{v_j} - 2$ because node v_j has $d_{v_j} - 1$ neighbors not in $\{v_1, v_2, \cdots, v_j\}$ and node v_j is removed from the set after it becomes infected. Therefore, we have

$$\left|\mathcal{N}\left((v_1, \cdots, v_j)\right)\right| = \left|\mathcal{N}\left((v_1, \cdots, v_{j-1})\right)\right| + d_{v_j} - 2.$$

Repeatedly applying the equation above, we have

$$\left|\mathcal{N}\left((v_1, \cdots, v_j)\right)\right| = d_{v_1} + \sum_{i=2}^{j}(d_{v_i} - 2),$$

which implies that for a permitted permutation $\mathbf{v}$, we have

$$\mathrm{Pr}\,(\mathbf{V} = \mathbf{v}) = \frac{1}{n} \frac{1}{d_{v_1}} \prod_{j=3}^{m} \frac{1}{d_{v_1} + \sum_{i=2}^{j-1}(d_{v_i} - 2)}.$$

Considering the example in Figure 3.2, we have

$$\mathrm{Pr}\,(\mathbf{V} = (a, b, c, d)) = \frac{1}{n} \frac{1}{d_{v_1}} \prod_{j=3}^{m} \frac{1}{d_{v_1} + \sum_{i=2}^{j-1}(d_{v_i} - 2)} = \frac{1}{10} \times \frac{1}{3} \times \frac{1}{3 + 3 - 2} \times \frac{1}{4 + 1 - 2}.$$

Denoting by $\mathcal{P}_u$ the set of permitted permutations starting from node v, the probability of having the infection subgraph g_i given u as the source, denoted by p_u, is

$$p_u = \Pr\left(\mathcal{O}|u\right) = \sum_{v \in \mathcal{P}_u} \frac{1}{n}\frac{1}{d_u} \prod_{j=3}^{m} \frac{1}{d_u + \sum_{i=2}^{j-1}(d_{v_i} - 2)}. \tag{3.3}$$

We further define $R_u = |\mathcal{P}_u|$, e.g., the number of permitted permutations starting from node v.

3.1.3   REGULAR TREES

We first consider a special case where the network is a regular tree. A d-regular tree is a tree graph where all nodes have the same degree d. On a d-regular tree, each permitted permutation occurs with the same probability

$$\Pr\left(\mathbf{V} = \mathbf{v}\right) = \frac{1}{n} \prod_{j=1}^{m-1} \frac{1}{dj - 2(j-1)}.$$

Therefore, the ML estimator of the source is the node that has the largest number of permitted permutations, in other words, with the largest R_u:

$$v^* \in \arg\max_u p_u = \arg\max_u R_u.$$

Shah and Zaman [2011] names R_u *rumor centrality* of node u, and establishes the following result.

Theorem 3.1   *On a regular tree network, the ML estimator of the diffusion source is the node with the maximum rumor centrality, called the rumor center.*   □

3.1.4   GENERAL TREE NETWORKS

Unfortunately, for general tree networks with heterogeneous node degrees, the probabilities of different permitted permutations are different. Therefore, to obtain the ML estimator, we need to compute the probability of each permitted permutation. However, the number of permitted permutations increases exponentially as the size of the infection subgraph increases. To overcome this issue, a simple heuristic has been proposed in Shah and Zaman [2011].

Shah and Zaman [2011] observed that starting from the same node, the probabilities of different permitted permutations are different, but the majority of them have similar probabilities. This leads to the following heuristic. First, the algorithm identifies the "typical" probability of a permitted permutation starting from a node, say node u, and assumes that the nodes were infected in an order that is consistent with the BFS tree from node u. Let $\mathbf{v}_u^*$ be a permitted permutation which is consistent with the BFS tree. Then, Equation (3.3) can be approximated by

$$p_u \approx \Pr(\mathbf{V} = \mathbf{v}_u^* | u) R_u,$$

and subsequently,

$$v^* \in \arg\max_u \Pr(\mathbf{V} = \mathbf{v}_u^* | u) R_u.$$

Note the the algorithm above balances the number of permitted permutations and the probability of these permutations.

Example from Shah and Zaman [2011]
Consider the example in Figure 3.3. Conditioned on node a being the source, the probability of a "typical" permitted permutation associated with the BFS tree (as shown in Figure 3.3) is

$$\left(\frac{1}{4}\right)^4.$$

Conditioned on node b being the source, the probability of a "typical" permitted permutation

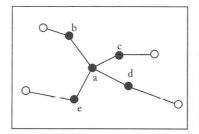 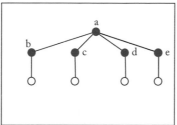 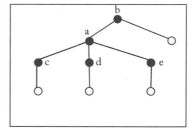

Figure 3.3: An example of rumor centrality on general networks.

associated with the BFS tree (as shown in Figure 3.3) is

$$\frac{1}{2}\left(\frac{1}{4}\right)^3.$$

Furthermore, it can be calculated that

$$R_a = 4!$$

because any permutation with node a as the first node is a permitted permutation in this case; and

$$R_b = 3!$$

because any permutation with b as the first node and a as the second node is a permitted permutation. Therefore,

$$\left(\frac{1}{4}\right)^4 \times 4! > \frac{1}{2}\left(\frac{1}{4}\right)^3 \times 3!,$$

and node a has a higher probability to be the source than node b.

Experiments results in Shah and Zaman [2011] demonstrated that this BFS heuristic improves the source localization accuracy over using the rumor center alone.

3.1.5   GENERAL NETWORKS

The source localization problem becomes even more difficult on general networks that are not trees. However, diffusion typically spreads over a spanning tree of a general network. Therefore, if the underlying infection spanning tree is given, the algorithm for the general tree networks can be applied to find a source estimator.

However, the underlying infection spanning tree itself is not easy to know. Shah and Zaman [2011] again proposed to use the BFS tree from a given infected node as the infection spanning tree, and then to calculate the value of $\Pr(V = v_u^*|u)R_u$ of the BFS tree, where v_u^* is a permitted permutation consistent with the BFS tree from node u. The source estimator is then chosen to be the node with the maximum $\Pr(V = v_u^*|u)R_u$.

We next discuss how to compute the rumor centrality of a node. Given a tree network, let $\mathcal{T}^v$ denote the tree rooted at node v and $\mathcal{T}_u^v$ denote the subtree of $\mathcal{T}^v$ rooted at node u. Furthermore, let T^v denote the number of nodes on $\mathcal{T}^v$, and T_u^v denote the number of nodes on subtree $\mathcal{T}_u^v$. Figure 3.4 shows a simple example. $T_b^a = 3$ since there are 3 nodes on the subtree rooted at node b. Similarly, we have $T_c^a = 1$.

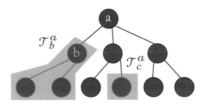

Figure 3.4: An example to illustrate $\mathcal{T}_u^v$ and T_u^v.

Next, we count the total number of permitted permutations R_v. Assume that the infection subgraph has m nodes, so each permutation is an ordered sequence of the m infected nodes. Since the first node must be the source node, node v is always the first node. Let k_v denote the number of infected neighbors of node v and child$(v) = \{u_1, u_2, \cdots, u_{k_v}\}$ denote the set of these infected neighbors. For the remaining $m - 1$ positions of the sequence, we divide these positions into k_v groups such that the jth group includes $T_{u_j}^v$ positions, which will be assigned to the nodes on $\mathcal{T}_{u_j}^v$. We will assign the positions to the subtrees one by one. After finishing assigning the positions to the first $j - 1$ subtrees, there are

$$\binom{m - 1 - \sum_{l=1}^{j-1} T_{u_l}^v}{T_{u_j}^v}$$

possible combinations for choosing the $T_{u_j}^v$ positions for the jth subtree. We therefore have

$$
R_v = \prod_{j=1}^{k_v} \binom{m-1-\sum_{l=1}^{j-1} T_{u_l}^v}{T_{u_j}^v} R_{u_j}(\mathcal{T}_{u_j}^v)
$$

$$
= \prod_{j=1}^{k_v} \frac{\left(m-1-\sum_{l=1}^{j-1} T_{u_l}^v\right)!}{T_{u_j}^v! \left(m-1-\sum_{l=1}^{j} T_{u_l}^v\right)!} R_{u_j}(\mathcal{T}_{u_j}^v)
$$

$$
= (m-1)! \prod_{u \in \text{child}(v)} \frac{R_u(\mathcal{T}_u^v)}{T_u^v!}, \tag{3.4}
$$

where $R_u(\mathcal{T})$ is the rumor centrality of node u on subgraph $\mathcal{T}$. Equation (3.4) is a recursive equation so we can expand it one step further and obtain

$$
R_v = (m-1)! \prod_{u \in \text{child}(v)} \frac{(T_u^v-1)! \prod_{w \in \text{child}(u)} \frac{R_w(\mathcal{T}_w^v)}{T_w^v!}}{T_u^v!}
$$

$$
= (m-1)! \prod_{u \in \text{child}(v)} \frac{1}{T_u^v} \prod_{w \in \text{child}(u)} \frac{R_w(\mathcal{T}_w^v)}{T_w^v!}. \tag{3.5}
$$

Note, a leaf node l only has 1 permitted permutation, so $R_l(\mathcal{T}_l^v) = 1$. The recursive equation will finally become

$$
R_v = (m-1)! \prod_{u:u \neq v} \frac{1}{T_u^v}. \tag{3.6}
$$

Since $T_v^v = m$, we have

$$
R_v = m! \prod_{u \in \mathcal{V}(g_i)} \frac{1}{T_u^v}. \tag{3.7}
$$

Example

Consider the example in Figure 3.4. Based on Equation (3.7), the rumor centrality of node a is

$$
R_a = \frac{10!}{10 \times 3 \times 3 \times 3} = 13{,}440.
$$

To calculate rumor centrality of all nodes in the infection subgraph, we need to calculate the size of subtree $\mathcal{T}_u^v$ for each pair of v and u, so the computational complexity is $O(n^2)$. This complexity can be reduced to $O(n)$ for tree networks by using a message-passing algorithm. We refer readers who are interested to Shah and Zaman [2011]

3.2 ACCURACY OF RUMOR CENTERS

This section analyzes the accuracy of rumor centrality on different tree graphs. We will see both positive and negative results. In particular, for a line graph, the asymptotic detection probability of using rumor center is zero, and for any regular tree which grows faster than a line, the detection probability is strictly positive.

3.2.1 LINE GRAPHS

In this section, we consider a line graph with infinitely many nodes and establish the following result.

Theorem 3.2 On a line graph, at time t, the probability that the rumor center is the actual source is $O\left(\frac{1}{\sqrt{t}}\right)$.

Proof. Denote by $\mathcal{D}_t$ the event that at time t the actual source coincides with the rumor center. We note that the diffusion process over a line network consists of two independent diffusion processes starting from the source. Let $K_l(t)$ denote the number of infected nodes on the left side of the source and $K_r(t)$ denote the number of infected nodes on the right side of the source. Both numbers are random variables. If the infection subgraph, which is also a line, has an odd number of nodes, the rumor center is the node in the middle. Otherwise, the two nodes in the middle are rumor centers, and the algorithm randomly chooses one as the source estimator. Therefore, when $K_l(t) = K_r(t)$, the unique rumor center is the source, and when $K_l(t)$ and $K_r(t)$ differs by 1, the actual source is found with probability 0.5. Therefore, we have

$$\Pr(\mathcal{D}_t) = \Pr(K_l(t) = K_r(t)) + \frac{1}{2}\Pr\left(|K_l(t) - K_r(t)| = 1\right).$$

Due to symmetry, $\Pr\left(K_l(t) = K_r(t) + 1\right) = \Pr\left(K_r(t) = K_l(t) + 1\right)$, so

$$\Pr(\mathcal{D}_t) = \Pr(K_l(t) = K_r(t)) + \Pr\left(K_l(t) = K_r(t) + 1\right) \tag{3.8}$$

$$= \sum_{k=1}^{\infty} \Pr(K_l(t) = K_r(t) = k) + \Pr(K_l(t) = k + 1, K_r(t) = k) \tag{3.9}$$

$$= \sum_{k=1}^{\infty} \left(\frac{t^k e^{-t}}{k!}\right)^2 + \frac{t^k e^{-t}}{k!}\frac{t^{k+1} e^{-t}}{(k+1)!}, \tag{3.10}$$

In the equation above, $\frac{t^k e^{-t}}{k!}$ is the probability of having exactly k infected nodes at time t at one side. This holds because the infection time is exponentially distributed so $K_l(t)$ is a Poisson random variable with mean t.

Define $g_k(t) = \left(e^{-t}\frac{t^k}{k!}\right)^2$. We next show that

$$\sum_{k=0}^{\infty} g_k(t) = O\left(\frac{1}{\sqrt{t}}\right).$$

Note that

$$\frac{g_k(t)}{g_{k-1}(t)} = \left(\frac{t}{k}\right)^2,$$

so $g_k(t)$ achieves its maximum value when $k = \lfloor t \rfloor$.

When $k \geq \lfloor t \rfloor$, we have

$$\frac{g_{k+\sqrt{t}}(t)}{g_k(t)} = \left(\frac{t^{\sqrt{t}}}{\prod_{j=1}^{\sqrt{t}}(k+j)}\right)^2 = \left(\frac{\prod_{j=1}^{\sqrt{t}} t}{\prod_{j=1}^{\sqrt{t}}(k+j)}\right)^2$$

$$= \prod_{j=1}^{\sqrt{t}}\left(\frac{t}{k+j}\right)^2 \leq \prod_{j=1}^{\sqrt{t}}\left(\frac{t}{t+j-1}\right)^2$$

$$= \prod_{j=1}^{\sqrt{t}}\left(1 + \frac{j-1}{t}\right)^{-2}$$

$$\leq \prod_{j=1}^{\sqrt{t}} e^{-\frac{j-1}{t}},$$

where the last inequality holds because $1 + x \geq e^{\frac{x}{2}}$ for $x \in [0,1]$. Note that

$$\sum_{j=1}^{\sqrt{t}} \frac{j-1}{t} = \frac{(\sqrt{t}-1)\sqrt{t}}{2t} = \frac{1}{2} - \frac{1}{2\sqrt{t}} \geq 13,$$

where the last inequality holds when $t \geq 36$. From the discussion above, we obtain

$$\frac{g_{k+\sqrt{t}}(t)}{g_k(t)} \leq e^{-\frac{1}{3}}.$$

Similarly, for $\lfloor t \rfloor - \sqrt{t} \leq k \leq \lfloor t \rfloor$, it can be obtained that

$$\frac{g_{k+\sqrt{t}}(t)}{g_k(t)} \geq e^{-\frac{1}{2}}.$$

Details can be found in Shah and Zaman [2011].

Now for $t \geq 36$, we have

$$\sum_{k=1}^{\infty} g_k(t) = \sum_{k=\lfloor t \rfloor}^{\lfloor t \rfloor + \sqrt{t} - 1} \left(\sum_{l=0}^{\infty} g_{k+l\sqrt{t}}(t) \right) + \sum_{k=\lfloor t \rfloor - \sqrt{t} + 1}^{\lfloor t \rfloor} \left(\sum_{l=0}^{\infty} g_{k-l\sqrt{t}}(t) \right) - g_{\lfloor t \rfloor},$$

where we define $g_k(t) = 0$ for $k \leq 0$ for convenience. Since $g_k(t)$ receives its maximum when $k = \lfloor t \rfloor$, we further have that for $k \geq \lfloor t \rfloor$

$$\sum_{l=0}^{\infty} g_{k+l\sqrt{t}}(t) \leq \sum_{l=0}^{\infty} g_k e^{-\frac{l}{3}} \leq g_{\lfloor t \rfloor} \frac{1}{1 - e^{-\frac{1}{3}}}.$$

Similarly, for $k \leq \lfloor t \rfloor$

$$\sum_{l=0}^{\infty} g_{k-l\sqrt{t}}(t) \leq \sum_{l=0}^{\infty} g_k e^{-\frac{l}{2}} \leq g_{\lfloor t \rfloor} \frac{1}{1 - e^{-\frac{1}{2}}}.$$

Therefore, for $t \geq 36$,

$$\sum_{k=1}^{\infty} g_k(t) \leq \sqrt{t} g_{\lfloor t \rfloor}(t) \left(\frac{1}{1 - e^{-1/3}} + \frac{1}{1 - e^{-1/2}} \right) = O\left(\sqrt{t} g_{\lfloor t \rfloor}(t) \right).$$

Finally, based on Stirling's approximation,

$$g_{\lfloor t \rfloor}(t) \propto \Theta\left(\frac{1}{\lfloor t \rfloor} \right),$$

which yields

$$\sum_{k=1}^{\infty} g_k(t) = O\left(\frac{1}{\sqrt{t}} \right).$$

Similarly, we can get

$$\sum_{k=1}^{\infty} \frac{t^k e^{-t}}{k!} \frac{t^{k+1} e^{-t}}{(k+1)!} = O\left(\frac{1}{\sqrt{t}} \right),$$

which concludes the proof. □

We also would like to remark that on a line graph, it has been shown in Spencer and Srikant [2015] that if the diffusion starts from two nodes instead of one, then not only the probability of the rumor source is asymptotically zero, but also the probability that the source is in any subgraph with size $o(N)$ approaches zero. In other words, even locating an $o(N)$ neighborhood of the sources is asymptotically impossible.

3.2.2 3-REGULAR TREES

Note a line graph is a regular tree with degree 2. It turns out that the detection probabilities using rumor centers are drastically different when the degree is greater than 2. It is shown in Shah and Zaman [2011] that the asymptotic detection probability is a positive constant between 0 and 0.5 when the degree of a regular tree is greater than 2, instead of 0. In this book, we show the result for regular trees with degree 3, where the exact asymptotic detection probability can be calculated and is equal to $\frac{1}{4}$. We next present the proof that shows that the asymptotic detection probability is at least $\frac{1}{4}$. The proof that $\frac{1}{4}$ is the upper bound is similar and can be found in Shah and Zaman [2011]. The proof is based on the following important property of rumor centers.

Lemma 3.3 *Assume the infection subtree has m nodes. Node v is a unique rumor center if*

$$T_u^v < \frac{m}{2} \quad \forall u \in child(v).$$

Proof. Suppose each of the subtrees rooted at the children of v has less than $\frac{m}{2}$ nodes. Then for any node $w \neq v$, we can always find a subtree $\mathcal{T}_x^w$ such that $T_x^w > \frac{m}{2}$, where x is a neighbor of w.

Recall that it has been shown that in Equation (3.6) that for any node w,

$$R_w = (m-1)! \prod_{u:u \neq w} \frac{1}{T_u^w}. \tag{3.11}$$

As shown in Figure 3.5, if x and w are neighbors, then

$$\mathcal{T}_y^w = \mathcal{T}_y^x \quad y \neq \{x, w\}.$$

Therefore, we have if x and w are neighbors, then

$$\frac{R_w}{R_x} = \frac{T_w^x}{T_x^w}.$$

From this equation, we can observe that $R_w < R_x$ if $T_w^x < T_x^w$, and in such case, node w is not the rumor center by definition.

Note that $T_w^x + T_x^w = n$. Therefore, if $T_x^w > n/2$, then $T_w^x < T_x^w$ and node w is not the rumor center. In conclusion, if each of the subtrees rooted at the children of v has less than $\frac{m}{2}$ nodes, then no other infected node is a rumor center, so node v is the unique rumor center. $\square$

Theorem 3.4 On a 3-regular tree, we have

$$\liminf_{t \to \infty} \Pr(\mathcal{D}_t) \geq \frac{1}{4}.$$

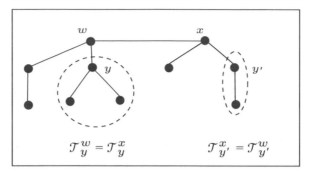

Figure 3.5: Example that shows $\mathcal{T}_y^w = \mathcal{T}_y^x$ $y \neq \{x, w\}$.

Proof. In a 3-regular tree, the source has three neighbors. Denote by $K_j(t)$ $(j = 1, 2, 3)$ the number of infected nodes on the subtree rooted the jth neighbor of the source. Since both the graph and diffusion process are symmetric, $K_j(t)$ are independent and identically distributed random variables. In the following analysis, we focus on $K_1(t)$ and drop subscript "1".

It is shown in Shah and Zaman [2011] that $K(t)$ is a geometric random variable with parameter e^{-t}. In other words, define $p = e^{-t}$, we have

$$\Pr(K(t) = m) = p(1 - p)^m.$$

Define $\mathcal{H}_t^m$ to be a set of tuples (m_1, m_2, m_3) such that each tuple in the set satisfies the following two conditions: (1) $m_1 + m_2 + m_3 = n$ and (2) $\max_j m_j < \frac{m}{2}$. From Lemma 3.3, we have

$$\bigcup_{m=0}^{\infty} \mathcal{H}_t^m \subset \mathcal{D}_t.$$

In other words, if the number of infected nodes on each of the three subtrees is less than half of the total number of infected nodes, then the rumor center is the diffusion source.

Therefore, we can lower bound $\Pr(\mathcal{D}_t)$ such that

$$\Pr(\mathcal{D}_t) \geq \sum_{m=0}^{\infty} \sum_{(m_1, m_2, m_3) \in \mathcal{H}_t^m} \prod_{j=1}^{3} \Pr(K_j(t) = m_j) \tag{3.12}$$

$$= \sum_{m=0}^{\infty} \sum_{(m_1, m_2, m_3) \in \mathcal{H}_t^m} p^3 (1 - p)^m \tag{3.13}$$

$$= p^3 \sum_{m=0}^{\infty} (1 - p)^m |\mathcal{H}_t^m|. \tag{3.14}$$

Now we calculate the size of $\mathcal{H}_t^m$. Suppose $m_1 > m_2 > m_3$. Then tuple $(m_1, m_2, m_3) \in \mathcal{H}_t^m$ needs to satisfy the following conditions:

$$\frac{m}{3} < m_1 < \frac{m}{2}$$
$$\frac{m - m_1}{3} < m_2 < m_1$$
$$m_1 + m_2 + m_3 = m.$$

Therefore, m_1 can take $\frac{m}{6} - 1$ different values, and given m_1, m_2 can take $\frac{3m_1 - n}{2} - 1$ different values. The total number of such tuples we have is

$$\sum_{m_1 = \frac{m}{3} - 1}^{\frac{m}{2} - 1} \frac{3m_1 - n}{2} - 1 = \frac{m^2}{48} + O(n).$$

Each permutation of (m_1, m_2, m_3) is a valid tuple in $\mathcal{H}_t^m$, so

$$|\mathcal{H}_t^m| \geq \frac{m^2}{8} + O(n).$$

Next we simplify Equation (3.14) and obtain that

$$\Pr(\mathcal{D}_t) \geq \frac{(1 - p)^2}{4} + O(p(1 - p) + p^2).$$

Recall that $p = e^{-t}$ as $t \to \infty$, $p \to 0$. Therefore, we have

$$\liminf_{t \to \infty} \Pr(\mathcal{D}_t) \geq \frac{1}{4}.$$

$\square$

Besides showing that the asymptotic detection probability is exactly 1/4 for 3-regular trees, Shah and Zaman [2011] also shows that the asymptotic detection probability is lower bounded by a constant α on any regular trees with degree at least tree.

The use of rumor centrality for source detection has been extended to other scenarios. Examples include multiple sources [Luo et al., 2013], single source with partial observations [Karamchandani and Franceschetti, 2013], single source with a priori distribution [Dong et al., 2013], and single source with multiple infection instances [Wang et al., 2014].

CHAPTER 4

Source Localization with Partial Timestamps

This chapter is devoted to source localization problem with partial timestamps, where besides the single snapshot of the network, the infection times of some infected nodes are also given to us. We first summarize the models used in this chapter.

- Network model: Same as the previous chapters, the network is a graph, which can be directed or undirected.

- Diffusion model: We assume general contact-based diffusion models under which a node can only be infected by its infected neighbors.

- Information model: The information includes $\mathcal{I}$ and $\boldsymbol{\tau}$, which is a $|\mathcal{I}|$-dimensional vector such that $\tau_v = \star$ if the timestamp is missing, and otherwise, τ_v is the time at which node v was infected.

We remark that the time considered in this chapter is the normal clock time, not the relative time with respect to the infection time of the source. This is important because estimating the relative infection time of the source is often equally difficult, if not more, than estimating the location of the source.

Figure 4.1 is a simple example showing the available information where the infection times of three nodes are given to us. Now given a spanning tree $\mathcal{T}$ on the infection subgraph, and a $|\mathcal{I}|$-dimensional vector $\mathbf{t}$ that specifies the time at which each infection occurs, we say $\mathbf{t}$ is *feasible* for spanning tree $\mathcal{T}$ if the infection time of a node is larger than that of its parent node on $\mathcal{T}$, and is *consistent* with the partial timestamps $\boldsymbol{\tau}$ if $t_v = \tau_v$ for all v such that $\tau_v \neq \star$. We call the tuple $(\mathcal{T}, \mathbf{t})$ a sample path, which defines both the direction and time infection occurred. Figure 4.2 shows a sample path that is feasible and consistent with the observation shown in Figure 4.1. Note that, for simplicity, we omitted the date in the figure by assuming all events occurred on the same day. The timestamps in black are the observed timestamps and the ones in blue are assigned by us. Define $\mathcal{F}(\mathcal{I}, \boldsymbol{\tau})$ to be the set of sample paths that are both feasible and consistent with the partial timestamps.

Given a sample path $\mathcal{P} = (\mathcal{T}, \mathbf{t}) \in \mathcal{F}(\mathcal{I}, \boldsymbol{\tau})$, we define the cost of the sample path to be

$$C(\mathcal{P}) = \sum_{(v,w) \in \mathcal{T}} (t_w - t_v - \mu)^2, \tag{4.1}$$

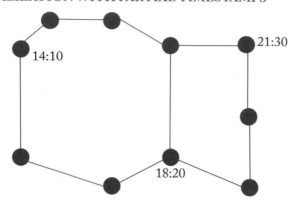

Figure 4.1: An example of available partial timestamps.

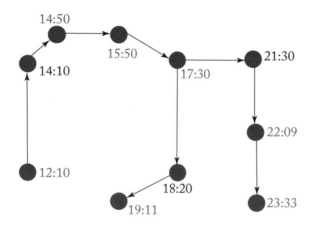

Figure 4.2: An example of a feasible and consistent sample path.

for some constant $\mu > 0$. This quadratic cost function is motivated by the following model [Pinto et al., 2012, Zhu et al., 2016]. Consider the continuous-time SI model but assume that the time it takes for node v to infect node w follows a truncated Gaussian distribution with mean μ and variance σ^2. Then given a sample path $\mathcal{P}$, the probability density associated with time sequence $\mathbf{t}$ is

$$f_{\mathcal{P}}(\mathbf{t}) = \prod_{(v,w)\in\mathcal{T}} \frac{1}{Z\sqrt{2\pi}\sigma} \exp\left(-\frac{(t_w - t_v - \mu)^2}{2\sigma^2}\right), \tag{4.2}$$

where Z is the normalization constant. Therefore, the log-likelihood is

$$\log f_{\mathcal{P}}(\mathbf{t}) = -(|\mathcal{I}| - 1)\log(Z\sqrt{2\pi}\sigma) - \frac{1}{2\sigma^2}\sum_{(v,w)\in\mathcal{T}}(t_w - t_v - \mu)^2,$$

where $|\mathcal{I}| - 1$ is the number of edges on tree $\mathcal{T}$. Therefore, given tree $\mathcal{T}$, a time sequence $\mathbf{t}$ is more likely to occur when the quadratic cost defined in (4.1) is smaller. Hence, identifying the most likely sample path (a sample-path-based estimator) is equivalent to finding the path with the lowest cost.

Now for a given infected node in the network, we define the cost associated with the node to be the minimum cost among all sample paths starting from the node. Using $\mathcal{P}_v$ to denote a sample path rooted at node v, the cost of node v is

$$C(v) = \min_{\mathcal{P}_v \in \mathcal{F}(\mathcal{I},\boldsymbol{\tau})} C(\mathcal{P}_v). \qquad (4.3)$$

After obtaining $C(v)$ for each infected node v, the sample-path-based estimator is the node v with the smallest $C(v)$. In other words, we again are using a sample-path-based approach for source localization.

Besides finding the sample-path-based estimator, we can further rank the infected nodes according to their likelihood of being the source. There are two heuristics that can be used to provide such a rank. We can either rank infected nodes according to $C(v)$ or according to the timestamps on the minimum cost sample path. We will see the performance of these two different heuristics later in this chapter.

We note that the calculation of $C(v)$ in a general graph is an NP-hard problem because the number of possible spanning trees is very large as stated in the following theorem. The proof of this theorem can be found in Zhu et al. [2016]. Therefore, we focus on heuristic algorithms.

Theorem 4.1   Problem (4.3) is an NP-hard problem.   □

4.1   EIF: A GREEDY ALGORITHM

To overcome the difficulty of problem (4.3), a greedy algorithm, named Earliest-Infection-First (EIF), has been proposed in Zhu et al. [2016]. The idea of EIF is to construct the sample path $(\mathcal{T}, \mathbf{t})$ by adding the infected nodes with observed infection times one by one. In particular, let us consider an example in Figure 4.3 (the same example used in Zhu et al. [2016]), where the figure on the left-hand-side is the observation we have. EIF iteratively constructs the sample path by adding infection nodes according to the observed infection times. In this example, five nodes have known infection times—nodes 10, 6, 12, 13, and 11 (listed in ascending order according to observed infection times). Now suppose at the beginning of current iteration, EIF has added nodes 10 and 6, and constructed a sample path

$$10 \rightarrow 6 \rightarrow 7 \rightarrow 8 \rightarrow 12. \qquad (4.4)$$

In Figure 4.3, the sample path is the subgraph formed by the red edges and the corresponding nodes. The edges on the current sample path are changed to directed edges as shown in Figure 4.3. We further assume that EIF assigns timestamps 6:45 PM to node 7 and 7:25 PM to node 8.

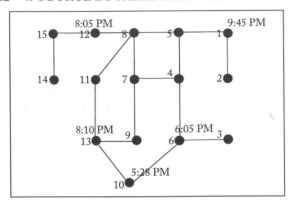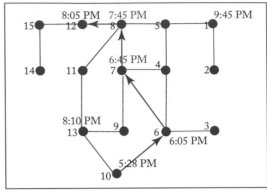

Figure 4.3: An example for illustrating EIF.

Now at the current iteration, EIF will add node 13, which has the *earliest observed infection time* among the infected nodes that have not been added to the sample path. To add node 13, EIF finds a path from node 13 to some node on the current sample path (4.4). To find the minimum cost path, EIF identifies the modified shortest path from each node on the current sample path to node 13, which is the path that connects the node to node 13 with the minimum number of hops without using any other nodes already on the current sample path and any node with observed infection time later than that of node 13. The modified shortest paths are

- Node 10 to node 13: $10 \rightarrow 13$.

- Node 6 to node 13: none

- Node 7 to node 13: $7 \rightarrow 9 \rightarrow 13$.

- Node 8 to node 13: $7 \rightarrow 11 \rightarrow 13$.

- Node 12 to node 13: none.

Note that there is no path from node 12 to node 13 because the edge between 8 and 12 is now a directed edge from 8–12. Also there is no modified shortest path from node 6 to node 13 because node 6 can only reach node 13 via node 7, which is already on the current sample path.

Now to calculate the quadratic cost of each of these paths, we apply the following lemma.

Lemma 4.2 *Consider a line with n infected nodes. Assume the infection times of node 1 and node n, which are two end nodes of the line, are known and the infection times of the other nodes are unknown. Furthermore, assume $\tau_1 < \tau_n$. The quadratic cost defined in (4.3) is minimized by setting*

$$t_k = \tau_1 + (k-1)\frac{\tau_n - \tau_1}{n-1} \tag{4.5}$$

for $1 < k < n$, *and the corresponding quadratic cost is*

$$\frac{(\tau_n - \tau_1 - \mu)^2}{n-1}.$$

$\square$

Proof. Define $x_{k,k-1} = t_k - t_{k-1}$, so the cost can be written as

$$C(\mathbf{x}) = \sum_{k=2}^{n} (t_k - t_{k-1} - \mu)^2 = \sum_{k=2}^{n} (x_{k,k-1} - \mu)^2.$$

The cost minimization problem can be written as

$$\min C(\mathbf{x}) = \sum_{k=2}^{n}(x_{k,k-1} - \mu)^2 \qquad (4.6)$$
$$\text{subject to:} \qquad \sum_{k=2}^{n} x_{k,k-1} = t_n - t_1 \qquad (4.7)$$
$$x_{k,k-1} \geq 0. \qquad (4.8)$$

Note that $C(\mathbf{x})$ is a convex function in $\mathbf{x}$. By verifying the KKT condition [Boyd and Vandenberghe, 2004], it can be shown that the optimal solution to the problem above is $x_{k,k-1} = \frac{\tau_n - \tau_1}{n-1}$, which implies $t_k = \tau_1 + (k-1)\frac{\tau_n - \tau_1}{n-1}$. $\square$

Note that under the optimal solution given in the previous lemma, the infection time, $\tau_{k+1} - \tau_k$, is the same for all edges, which is due to the quadratic form of the cost function. Given the example in Figure 4.3, the quadratic costs of the three modified shortest paths are

- Cost of $10 \to 13 : 15,640$

- Cost of $7 \to 9 \to 13 : 61.83$

- Cost of $8 \to 11 \to 13.\ 147.03$

Therefore, $7 \to 9 \to 13$ has the smallest cost. EIF adds path $7 \to 9 \to 13$ to the sample path and assigns timestamp 7:28 PM to node 9 as shown in Figure 4.4.

We now present a complete description of EIF from Zhu et al. [2016].

Earliest-Infection-First (EIF)

1. Step 1: The algorithm first estimates μ from τ using the average per-hop infection time. Let l_{vw} denote the length of the shortest path from node v to node w, then

$$\mu = \frac{\sum_{\tau_v \neq \star, \tau_w \neq \star, v \neq w} |\tau_v - \tau_w|}{\sum_{\tau_v \neq \star, \tau_w \neq \star, v \neq w} d_{vw}}.$$

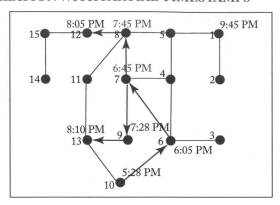

Figure 4.4: Adding node 13 to the sample path under EIF.

2. Step 2: Sort the infected nodes in ascending order according to the observed infection times τ. Let α denote the ordered list such that α_1 is the node with the earliest infection time.

3. Step 3: Construct the initial sample path $\mathcal{T}_0$ that includes the root node only and sets the cost to be zero.

4. Step 4: At the kth iteration, node α_k is added to the sample path $\mathcal{T}_{k-1}$ using the following steps.

   (a) For each node m on the sample path $\mathcal{T}_{k-1}$, identify a modified shortest path from node m to node α_k. The modified shortest path is a path that has the minimum number of hops among all paths from node m to node α_k, which satisfies the following two conditions:

   – it does not include any nodes on the sample path $\mathcal{T}_{k-1}$, except node m; and
   – it does not include any nodes on list α, except node α_k.

   (b) For the modified shortest path from node m to node α_k, the cost of the path is defined to be

   $$\gamma_m = \tilde{l}_{\alpha_k m} \left(\frac{t_{\alpha_k} - t_m}{\tilde{l}_{\alpha_k m}} - \mu \right)^2,$$

   where $\tilde{l}_{\alpha_k m}$ is the length of the modified shortest path from m to α_k. From all nodes on the sample path $\mathcal{T}_{k-1}$, select node m^* with the minimum cost, i.e.,

   $$m^* = \arg\min_m \gamma_m.$$

(c) Construct a new sample path $\mathcal{T}_k$ by adding the modified shortest path from m^* to α_k. Assume node g on the newly added path is h_g hops from node m^*; the infection time of node g is set to be

$$t_g = t_{m^*} + (h_g - 1)\frac{t_{\alpha_k} - t_{m^*}}{\tilde{l}_{m^*\alpha_k}}. \tag{4.9}$$

The cost is updated to $C(v) = C(v) + \gamma_{m^*}$.

5. Step 5: For those infected nodes that have not been added to the sample path, add these nodes by using a breadth-first search starting from the current sample path $\mathcal{T}$. When a new node (say node w) is added to the sample path during the breadth-first search, the infection time of the node is set to be $t_{p_w} + \mu$, where p_w is the parent of node w on the sample path. Note that the cost $C(v)$ does not change during this step because $t_w - t_{p_w} - \mu = 0$.

4.1.1 COST-BASED AND TREE-BASED RANKING

Denote by $\mathcal{T}_v$ the sample path constructed under EIF for node v, and $C(\mathcal{T}_v)$ the corresponding cost computed by EIF. After constructing the sample path for each infected node and obtaining the corresponding cost, the nodes are ranked using the following two approaches.

Cost-Based Ranking (CR): Rank the infected nodes in an ascending order according to $C(\mathcal{T}_v)$.

Tree-Based Ranking (TR): Denote by $v^* = \arg\min_v C(\mathcal{T}_v)$. Rank the infected nodes in an ascendent order according to the timestamps on $\mathcal{T}_{v^*}$.

4.2 GAUSSIAN HEURISTIC

The EIF algorithm is a greedy algorithm to reconstruct a sample path with minimum quadratic cost. A different heuristic has also been used earlier in Pinto et al. [2012], which is one of the first papers that studied source localization with partial timestamps. The heuristic in Pinto et al. [2012] uses the intuition that if the diffusion spreads very fast, then it likely spreads along the BFS tree. Therefore, for each node, the algorithm first constructs the BFS tree and then calculates the probability of having the observed infection times when node v is the source by restricting the diffusion to the BFS tree. Pinto et al. [2012] assumes the infection times follow Gaussian distributions. The estimator of the source is then chosen to be the node associated with the highest probability. We note that when the source is determined, we can also rank the nodes according to their positions on the corresponding BFS tree.

4.3   PERFORMANCE EVALUATION

In Zhu et al. [2016], we tested the algorithms using both synthetic data and real social network data. We present some of the simulation results in this book. Readers who are interested can find more results in Zhu et al. [2016].

For performance evaluation with synthetic data, we used two real-world networks, the Internet Autonomous Systems network (IAS) and the power grid network (PG), and compared with the following four algorithms.

- Rumor centrality (RUM): We can use RUM to rank the infected nodes in an ascending order according to nodes' rumor centrality.

- Infection eccentricity (ECCE): The infection eccentricity of a node is the maximum distance from the node to any infected node in the graph, where the distance is defined to be the length of the shortest path. Recall that the node with the smallest infection eccentricity, the Jordan infection center, is the optimal sample-path-based estimator as discussed in Chapter 2. ECCE ranks the infected nodes in descending order according to infection eccentricity.

- NETSLEUTH: NETSLEUTH can be used to rank nodes according to the eigenvector corresponding to the largest eigenvalue of the submatrix.

- Gaussian heuristic (GAU).

In the four algorithms above, only RUM, ECCE, and NETSLEUTH use topological information of the network and do not exploit the timestamp information. GAU utilizes partial timestamp information.

In this set of experiments, it is assumed the infection time of each infection follows a truncated Gaussian distribution with $\mu = 100$ and $\sigma = 100$. We selected 50% infected nodes (100 nodes) and revealed their infection time. The source node was always excluded from these 100 nodes so that the infection time of the source node was always unknown. We repeated the simulation 500 times to compute the average $\gamma\%$-accuracy.

The results on the IAS and PG networks are presented in Figures 4.5 and 4.6, respectively. We can observe that in both networks, CR and TR perform much better than the other algorithms. In particular, in the IAS network, the 10%-accuracy of CR is 0.76 while 10%-accuracy of GAU and NETSLEUTH is 0.48 and 0.43, respectively. In the PG network, the 10%-accuracy of TR is 0.99 while that of GAU and NETSLEUTH is 0.93 and 0.43, respectively.

Furthermore, most algorithms, except NETSLEUTH, have higher $\gamma\%$-accuracy in the PG network than that in the IAS network. We conjecture that it is because the IAS network has a

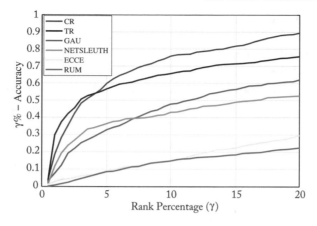

Figure 4.5: Comparison with existing algorithms in the IAS network with 50% timestamps.

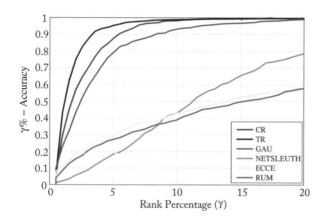

Figure 4.6: Comparison with existing algorithms in the PG network with 50% timestamps.

small diameter and contains hub nodes while the PG network is more tree-like. NETSLEUTH dominates ECCE and RUM in the IAS network but performs worse than ECCE and RUM in the PG network when $\gamma \leq 10$. Furthermore, while all other algorithms have higher γ-accuracy in IAS than in PG, NETSLEUTH has lower γ-accuracy in IAS than in PG when $\gamma < 10$. A similar phenomenon will be observed in a later simulation as well.

Weibo Data Evaluation
Zhu et al. [2016] also evaluated the performance of our algorithms with real-world network and real-world information spreading. The dataset is the Sina Weibo[1] data, provided by the WISE

[1]http://www.weibo.com/

2012 challenge.[2] Sina Weibo is the Chinese version of Twitter. The dataset includes a friendship graph and a set of tweets.

The friendship graph is a directed graph with 265,580,802 edges and 58,655,849 nodes. The tweet dataset includes 369,797,719 tweets. Each tweet includes the user ID and post time of the tweet. If the tweet is a retweet, it includes the tweet ID of the original tweet, the user who posted the original tweet, the post time of the original tweet, and the retweet path of the tweet which is a sequence of user IDs. For example, the retweet path $a \to b \to c$ means that user b retweeted user a's tweet, and user c retweeted user b's.

We selected the tweets with more than 1,500 retweets. For each tweet, all users who retweet the tweet are viewed as infected nodes and we extracted the subnetwork induced by these users. We also added those edges on the retweet paths to the subnetwork if they are not present in the friendship graph by treating them as missing edges in the friendship network. The user who posts the original tweet is regarded as the source. If there does not exist a path from the source to an infected node along which the post time is increasing, the node was removed from the subnetwork.

Note that in some cases, the source can be easily located using a naive algorithm, e.g., when the network is star or only the source can reach all other infected nodes. To avoid these cases, we further selected the tweets that satisfy the following conditions:

- The number of infected nodes is at least 100.

- The diameter of the undirected version of the subnetwork is at least 7.

- There exist at least 50 nodes that can reach all other infected nodes in the network.

- At least 30% of nodes have timestamps. This is to make sure we have enough timestamps for evaluating CR and TR.

After removing tweets that do not satisfy the above conditions, we have 347 tweets with at least 30% observed timestamps. The $\gamma\%$-accuracy is summarized in Figure 4.7, where we include the results with 10% of timestamps and 30% of timestamps. The observed timestamps are uniformly selected from the available timestamps and the source node is excluded. RUM and ECCE are not included in the figure since the performance is dominated by NETSLEUTH.

The figure shows that CR and TR dominate GAU with both 10% and 30% of timestamps. In particular, TR performs very well and dominates all other algorithms with a large margin. The 10%-accuracy of TR with 30% timestamps is around 0.81 while that of CR is 0.53 and that of NETSLEUTH is only 0.39.

[2]http://www.wise2012.cs.ucy.ac.cy/challenge.html

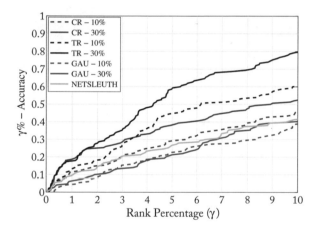

Figure 4.7: Performance on Weibo data.

CHAPTER 5

Open Questions

While significant progress has been made on source localization in large-scale networks under various different models, a number of questions remain open. We next summarize the open questions in three categories: (1) more general diffusion models, (2) realistic networks, and (3) incomplete observations.

- **Source localization beyond contact-based diffusion models.** The current results focused on contact-based diffusion models such as the IC model or SIR model. Another class of popular diffusion models is strategic diffusion models such as linear threshold model [Granovetter, 1978, Kempe et al., 2003, Schelling, 1978], the exposure-curve model [Myers et al., 2012, Romero et al., 2011], and the Ising model [Liu et al., 2010], where a node strategically involves in the diffusion process based on the states of its neighbors. Both fundamental limits and effective algorithms are likely to be very different under strategic diffusion models. In particular, the following questions need to be addressed:

  - How do strategic behaviors change diffusion processes and impact source localization?

  - If a diffusion process starts from multiple sources, can diffusion processes starting from different sources be separated (at least partially)? How to estimate the number of sources and further locate them?

- **Source localization in realistic networks.** While the ER random graph [Erdos and Renyi, 1959, 1960] exhibits the small world property, it is not a power-law network. It is important to study source localization in realistic graph models such as random dot product graph [Nickel, 2006, Young and Scheinerman, 2007] or Kronecker graphs [Leskovec et al., 2010], which produce networks with small diameters, power-law degree distributions, triangle participation, etc. It would also be interesting to explore source localization in spatial networks such as random geometric graphs, where two nodes are connected if they are close in physical proximity. In particular, the following questions remain to be addressed:

  - How do small diameters, power-law, triangle participation affect the rate of diffusion and the accuracy of source localization? Can additional graph characteristics such as community structures be utilized to improve source localization accuracy?

  - How can we extend the results to spatial networks?

– How can we design effective and scalable source detection algorithms that apply to all the above general and realistic graph models?

• **Source localization with incomplete observations**

Source localization problem has been studied under various different assumptions on the available information, including partial observation, partial timestamps, etc. Theoretical results have been established only for tree networks. However, the fundamental tradeoff between information and localization accuracy remains open.

– Assuming no timestamp information, what is the fundamental tradeoff between the sampling rate (the probability each node is observed) and source localization accuracy?

– What are the fundamental impacts of the size of timestamps and its distribution?

Bibliography

http://www.cdc.gov/flu/spotlights/pandemic-global-estimates.htm 1

A.-L. Barabasi and R. Albert. Emergence of scaling in random networks. *Science*, 286(5439), pages 509–512, 1999. DOI: 10.1515/9781400841356.349. 2

S. Boyd and L. Vandenberghe. *Convex Optimization*. Cambridge University Press, New York, 2004. DOI: 10.1017/cbo9780511804441. 63

W. Chen, Y. Wang, and S. Yang. Efficient influence maximization in social networks. In *Proc. Annual ACM SIGKDD Conference Knowledge Discovery and Data Mining (KDD)*, pages 199–208, 2009. DOI: 10.1145/1557019.1557047. 2

Z. Chen, K. Zhu, and L. Ying. Detecting multiple information sources in networks under the SIR model. *IEEE Transactions on Network Science Engineering*, 3(1), pages 17–31, 2016. DOI: 10.1109/tnse.2016.2523804. 44

W. Dong, W. Zhang, and C. W. Tan. Rooting out the rumor culprit from suspects. In *Proc. IEEE International Symposium Information Theory (ISIT)*, pages 2671–2675, Istanbul, Turkey, 2013. DOI: 10.1109/isit.2013.6620711. 57

V. M. Eguiluz and K. Klemm. Epidemic threshold in structured scale-free networks. *Physical Review Letters*, 89(10):108701, September 2002. DOI: 10.1103/PhysRevLett.89.108701. 2

P. Erdos and A. Renyi. On random graphs I. *Publicationes Mathematicae Debrecen*, 6, pages 290–297, 1959. 1, 2, 17, 71

P. Erdos and A. Renyi. *On the Evolution of Random Graphs*. Akad. Kiado, 1960. 2, 71

S. Feizi, K. Duffy, M. Kellis, and M. Medard. Network infusion to infer information sources in networks. *MIT-CSAIL-TR-2014-028, Technical Report*, 2014. 2

A. Ganesh, L. Massoulie, and D. Towsley. The effect of network topology on the spread of epidemics. In *Proc. IEEE International Conference Computer Communications (INFOCOM)*, pages 1455–1466, Miami, March 2005. DOI: 10.1109/infcom.2005.1498374. 2

J. Goldenberg, B. Libai, and E. Muller. Talk of the network: A complex systems look at the underlying process of word-of-mouth. *Marketing Letters*, 12(3), pages 211–223, 2001. DOI: 10.1023/A:1011122126881. 1, 5

74 BIBLIOGRAPHY

M. Granovetter. Threshold models of collective behavior. *American Journal of Sociology*, 83(6), pages 1420–1443, May 1978. DOI: 10.1086/226707. 71

D. Gruhl, R. Guha, D. Liben-Nowell, and A. Tomkins. Information diffusion through blogspace. In *Proc. of the International Conference World Wide Web (WWW)*, pages 491–501, New York, 2004. DOI: 10.1145/988672.988739. 2

F. Ji, W. P. Tay, and L. R. Varshney. Estimating the number of infection sources in a tree. In *Proc. IEEE Global Conference on Signal and Information Processing (GlobalSIP)*, pages 380–384, 2016. DOI: 10.1109/globalsip.2016.7905868. 44

F. Ji, W. P. Tay, and L. Varshney. An algorithmic framework for estimating rumor sources with different start times. *IEEE Transactions on Signal Processing*, 65(10), pages 2517–2530, 2017. DOI: 10.1109/tsp.2017.2659643. 44

N. Karamchandani and M. Franceschetti. Rumor source detection under probabilistic sampling. In *Proc. IEEE International Symposium Information Theory (ISIT)*, Istanbul, Turkey, July 2013. DOI: 10.1109/isit.2013.6620613. 57

D. Kempe, J. Kleinberg, and E. Tardos. Maximizing the spread of influence through a social network. In *Proc. Annual ACM SIGKDD Conference Knowledge Discovery and Data Mining (KDD)*, pages 137–146, Washington DC, 2003. DOI: 10.1145/956755.956769. 2, 10, 71

D. Kempe, J. Kleinberg, and E. Tardos. Influential nodes in a diffusion model for social networks. *Automata, Languages and Programming*, pages 1127–1138, 2005. DOI: 10.1007/11523468_91. 2

J. O. Kephart and S. R. White. Directed-graph epidemiological models of computer viruses. In *IEEE Computer Society Symposium Research in Security and Privacy*, pages 343–359, Orkland, CA, 1991. DOI: 10.1109/risp.1991.130801. 2

J. Leskovec, D. Chakrabarti, J. Kleinberg, C. Faloutsos, and Z. Ghahramani. Kronecker graphs: An approach to modeling networks. *Journal of Machine Learning Research*, 11, pages 985–1042, March 2010. 71

S. Liu, L. Ying, and S. Shakkottai. Influence maximization in social networks: An Ising-model-based approach. In *Proc. Annual Allerton Conference Communication, Control and Computing*, pages 570–576, Monticello, IL, 2010. DOI: 10.1109/allerton.2010.5706958. 71

W. Luo, W. P. Tay, and M. Leng. On the universality of Jordan centers for estimating infection sources in tree networks. *IEEE Transactions on Information Theory*, 63(7), pages 4634–4657, July 2017. DOI: 10.1109/tit.2017.2698504. 25, 32

W. Luo, W. P. Tay, and M. Leng. Identifying infection sources and regions in large networks. *IEEE Transactions on Signal Processing*, 61, pages 2850–2865, 2013. DOI: 10.1109/tsp.2013.2256902. 44, 57

C. R. MacCluer. The many proofs and applications of Perron's theorem. *Siam Review*, 42(3), pages 487–498, 2000. DOI: 10.1137/s0036144599359449. 35

A. Medina, I. Matta, and J. Byers. On the origin of power laws in Internet topologies. *ACM SIGCOMM Computer Communication Review*, 30(2), pages 18–28, 2000. DOI: 10.1145/505680.505683. 2

C. Moore and M. E. J. Newman. Epidemics and percolation in small-world networks. *Physical Review E*, 61(5), pages 5678–5682, 2000. DOI: 10.1103/physreve.61.5678. 2

S. A. Myers, C. Zhu, and J. Leskovec. Information diffusion and external influence in networks. In *Proc. Annual ACM SIGKDD Conference Knowledge Discovery and Data Mining (KDD)*, pages 33–41, Beijing, China, 2012. DOI: 10.1145/2339530.2339540. 71

M. E. J. Newman. The spread of epidemic disease on networks. *Physical Review E*, 66(1):016128, July 2002. DOI: 10.1103/physreve.66.016128. 2

M. E. J. Newman and D. J. Watts. Renormalization group analysis of the small-world network model. *Physical Letters A*, 263(4), pages 341–346, 1999. DOI: 10.1016/s0375-9601(99)00757-4. 2

M. E. J. Newman, S. Forrest, and J. Balthrop. Email networks and the spread of computer viruses. *Physical Review E*, 66(3):035101, September 2002. DOI: 10.1103/physreve.66.035101. 2

C. L. M. Nickel. *Random Dot Product Graphs: A Model for Social Networks*. Ph.D. thesis, Johns Hopkins University, 2006. 2, 71

R. Pastor-Satorras and A. Vespignani. Epidemic spreading in scale-free networks. *Physical Review Letters*, 86(14), pages 3200–3203, 2001. DOI: 10.1103/physrevlett.86.3200. 2

P. C. Pinto, P. Thiran, and M. Vetterli. Locating the source of diffusion in large-scale networks. *Physical Review Letters*, 109(6):068702, 2012. DOI: 10.1103/physrevlett.109.068702. 60, 65

B. A. Prakash, J. Vreeken, and C. Faloutsos. Spotting culprits in epidemics: How many and which ones? In *IEEE International Conference Data Mining (ICDM)*, pages 11–20, Brussels, Belgium, 2012. DOI: 10.1109/icdm.2012.136. 32, 36, 44

D. M. Romero, B. Meeder, and J. Kleinberg. Differences in the mechanics of information diffusion across topics: Idioms, political hashtags, and complex contagion on twitter. In *Proc. International Conference World Wide Web (WWW)*, pages 695–704, Hyderabad, India, 2011. DOI: 10.1145/1963405.1963503. 71

T. C. Schelling. *Micromotives and Macrobehavior*. Norton, 1978. 71

D. Shah and T. Zaman. Rumors in a network: Who's the culprit? *IEEE Transactions on Information Theory*, 57, pages 5163–5181, August 2011. DOI: 10.1109/tit.2011.2158885. 37, 45, 46, 48, 49, 50, 51, 53, 55, 56, 57

S. Spencer and R. Srikant. On the impossibility of localizing multiple rumor sources in a line graph. *ACM SIGMETRICS Performance Evaluation Review*, 43(2), pages 66–68, 2015. DOI: 10.1145/2825236.2825262. 54

Z. Wang, W. Dong, W. Zhang, and C. W. Tan. Rumor source detection with multiple observations: Fundamental limits and algorithms. In *Proc. Annual ACM SIGMETRICS Conference*, Austin, TX, 2014. DOI: 10.1145/2637364.2591993. 57

D. J. Watts and S. H. Strogatz. Collective dynamics of small-world networks. *Nature*, 393(6684), pages 440–442, 1998. DOI: 10.1515/9781400841356.301. 2

S. J. Young and E. R. Scheinerman. Random dot product graph models for social networks. In *Proc. International Conference Algorithms and Models for the Web-Graph*, pages 138–149, San Diego, CA, 2007. DOI: 10.1007/978-3-540-77004-6_11. 71

K. Zhu and L. Ying. Information source detection in the SIR model: A sample path based approach. In *Proc. Information Theory and Applications Workshop (ITA)*, February 2013a. DOI: 10.1109/ita.2013.6502991. 25, 26, 27

K. Zhu and L. Ying. A robust information source estimator with sparse observations. *Arxiv Preprint arXiv:1309.4846*, 2013b. https://link.springer.com/article/10.1186/s40649-014-0003-2 DOI: 10.1186/s40649-014-0003-2. 32

K. Zhu and L. Ying. Source localization in networks: Trees and beyond. *ArXiv:1510.01814*, 2015. 17, 18, 20, 37

K. Zhu and L. Ying. Information source detection in networks: Possibility and impossibility results. In *Proc. IEEE International Conference Computer Communications (INFOCOM)*, San Francisco, CA, 2016a. DOI: 10.1109/infocom.2016.7524363. 6, 7

K. Zhu and L. Ying. Information source detection in the sir model: A sample-path-based approach. *IEEE/ACM Transactions on Networks*, 24(1), pages 408–421, 2016b. DOI: 10.1109/tnet.2014.2364972. 25, 27, 28, 31

K. Zhu, Z. Chen, and L. Ying. Locating the contagion source in networks with partial timestamps. *Data Mining and Knowledge Discovery*, 30(5), pages 1217–1248, 2016. http://dx.doi.org/10.1007/s10618-015-0435-9 DOI: 10.1007/s10618-015-0435-9. 60, 61, 63, 66, 67

K. Zhu, Z. Chen, and L. Ying. Catch'Em All: Locating multiple diffusion sources in networks with partial observations. In *AAAI Conference Artificial Intelligence*, 2017. 44

Authors' Biographies

LEI YING

Lei Ying received his B.E. degree from Tsinghua University, Beijing, China, and his M.S. and Ph.D. in Electrical and Computer Engineering from the University of Illinois at Urbana-Champaign. He currently is an Associate Professor at the School of Electrical, Computer and Energy Engineering at Arizona State University.

His research interest is broadly in the area of stochastic networks, including cloud computing, communication networks, and social networks. He is coauthor with R. Srikant of the book *Communication Networks: An Optimization, Control and Stochastic Networks Perspective,* Cambridge University Press, 2014. He won the Young Investigator Award from the Defense Threat Reduction Agency (DTRA) in 2009 and NSF CAREER Award in 2010. He was the Northrop Grumman Assistant Professor in the Department of Electrical and Computer Engineering at Iowa State University from 2010 to 2012. His papers have received the best paper award at IEEE INFOCOM 2015 and the Kenneth C. Sevcik Outstanding Student Paper Award at ACM SIGMETRICS/IFIP Performance 2016, been selected in the ACM TKDD Special Issue "Best Papers of KDD 2016," received the WiOpt'18 Best Student Paper Award, and selected for Fast-Track Review for TNSE at IEEE INFOCOM 2018 (7 out of 312 accepted papers were invited).

KAI ZHU

Kai Zhu received his B.E. degree in Electronics Engineering from Tsinghua University, Beijing, China, in 2010 and his Ph.D. in Electrical Engineering from Arizona State University in 2015. His research interest is in social networks and data mining.

<barcode>‖‖ ‖ ‖‖ ‖‖‖‖‖ ‖‖‖ ‖ ‖‖‖‖‖‖ ‖‖‖ ‖‖‖‖ ‖‖ ‖‖ ‖‖‖‖ ‖‖‖ ‖‖ ‖‖‖ ‖‖‖ ‖‖‖</barcode>

Printed in the United States
by Baker & Taylor Publisher Services